EVERYONE'S FIELD GUIDE TO PHYSICS

Volume I: The Familiar World

By Frank Maybusher

DEDICATION

Dedicated to Hye-Jung, Isaac, and Aeon,

all of whom give meaning to this beautiful and mysterious universe.

TABLE OF CONTENTS

PREFACE

After brainstorming for many days on how to approach writing a book based upon the world of physics, I had to ask many questions. Who are my readers going to be? Young, old, students, casual readers, or those more in touch with the concepts of science? I had initially thought of writing on these topics with my two children in mind. My two sons aged nine and eleven, as of this writing, could benefit from dad's knowledge of the subject and may even learn to enjoy it! Then there were my parents, who may not have very much appreciation for it at all and may not have any idea as to its full and overwhelming power and influence in our daily lives. How about my wife, who *does* have an appreciation and a background touching on the sciences? What perspectives can I bring to her that she may not know? How may I make these particular subjects within the sciences more respectful and clearer to her?

So then, assuming I can settle on my audience's potential make-up, the next concern is how much detail should I pursue? How much history? The science of physics and its history has been detailed in an untold number of books throughout the ages and for all audiences. I can't pretend to break any ground here.

Every one of us, however, brings to life their own unique perspectives on the world. How one person sees the world may be entirely different from another, who may be witnessing the same thing. This perspective also changes with a single individual throughout their life as time marches on.

And so, it is with these subjects as well. Their descriptions of the universe and our world have changed throughout history and are constantly shifting. Physics concepts change and build upon earlier ones; adjusting, shifting, merging, and accumulating.

Then there is the big question that cannot at all be avoided when it comes to the 'hard' science of physics: *how much math should I know to understand them, and should there be any discussed in the writing of it?*

The answer to this question is a clear and decisive, "Yes! You should be familiar with a little math and there will be some in this book...but!" Please be patient and let me explain.

Mathematics is the tool and the language used in understanding and explaining the universe. There really is nothing else that is used that has such an affinity for this science, or for that matter, any of the other physical sciences and engineering. Discussing these sciences without the help of mathematics would be worse than attempting to understanding the nuanced dialogues of Shakespeare's characters without the use of the English language. Yes, you can attempt to explain the interactions of Hamlet and Horatio without the use of English. However, you will never understand the full nuances nor the feelings that exist between these characters when they speak to one another, without the use of English. Yes, you can attempt to explain the visual world around us to the best of your ability, but to a person blind at birth the attempt will, for all intents and purposes, be in vain when discussing the feeling one has or experiences when seeing the difference between, say, a red sunset and a blue tropical sea.

However, within this book, I understand that my audience may want nothing to do with math. I understand that there are those who run in the opposite direction when the "m-word" is used. These are the people who prioritize their list of fears in life as "#2. Fear of death...#1. Fear of mathematics". For this group of reader (and for all that I know, it may be just about everyone reading this book), I have set aside those sections in the writing that are more mathematical. With a couple of exceptions, that I felt were crucial for inclusion within the general reading, these sections are shaded gray and are surrounded by a border. So, you may skip these areas of reading without missing the main ideas that I am trying to explain. For those more adventurous, more inclined, or more mathematically adept, you may read these areas and hopefully absorb a deeper understanding of the physics.

I have not grayed out, nor bordered, any of the mathematical discussions within any of the appendices, however. The appendices of the book are entirely optional reading. They are admittedly fascinating topics, though, and help explain other ideas that some of us may be curious about.

The "Taoist" side of my brain has settled on just explaining these topics as casually as I feel comfortable, and in as much detail as I feel like pursuing to an audience that I would label as friends who may have a touching appreciation for science in general.

If some of my audience grows bored or does not understand some aspects of my writing, please stick with it, everyone should be able to get this stuff! The whole idea of writing this book is to attempt to explain the beauty of physics with clarity and interest. I will try my best.

As for you, dear reader, welcome to the familiar world of physics!

INTRODUCTION

What is Physics?

"All science is either physics or stamp collecting."

- Ernest Rutherford

Physics is only one field of science, among many others. It is, however, the most fundamental of all sciences. All others arise and derive from it. Strictly speaking, physics investigates the interactions between matter, energy, and forces. But this definition doesn't do the subject justice. Physics wishes to delve as deep and as wide as possible within our universe to answer some of the most significant questions. What is everything made of? How big is our universe? How small can things get? What is time? Where did everything come from? How old is the universe? Where do we fit in to the universe? These questions and thousands more can be asked, but it is only through the field of physics that they can be answered. While many of these queries may overlap with other areas of science, such as biology and astronomy, they all share the underlying physics. Our curiosities inevitably bring us to this one science.

To have a greater appreciation for our place in the universe and where we come from, it helps to understand how nature works. Physics is a way of doing this. Without having some of this knowledge, we are left walking blind and ignorant through our lives. We miss out understanding the most important concepts that humanity has ever discovered.

Why should we be familiar with physics?

"Heaven wheels above you

Displaying to you her eternal glories,

But still your eyes are on the ground"

- Dante Alighieri, *The Divine Comedy*

We may all go through life concerned only with our daily chores, dealing with family issues, and enjoying ourselves and each other's company on vacations, at work, at school, etc. We watch the news to get the latest developments in the next state over, another country on the opposite side of the planet, oblivious to the fact that many of these concerns do not at all impact our day to day lives. Countless hours, from every walk of life and from all corners of the globe, are spent on social networks, getting the scoop on whatever exposure our friends choose to show to the world.

Humans are an inquisitive and social species and, of course, if given a chance to look through the peephole of social networking at each other, or the dirty laundry of news, we will look. Much of the hours of our lives may be filled with other concerns, like work. While our jobs may give us a sense of impact on the world and a sense pride for jobs well done, there may still be an occasional glimmer of something missing in our lives that we can't quite put our hands on.

Besides our daily work, we may fill the rest of our times with being close to our families and friends. We socialize, criticize, care, and support one another. But still, even after devoting our entire days to work, family and friends, a subtle

11

hole remains. That hole is in the shape of curiosity. But even then, I am reminded of a quote from the late wheelchair-bound physicist Stephen Hawking, who said:

> *"It surprises me how disinterested we are today about things like physics, space, the universe and philosophy of our existence, our purpose, our final destination. It's a crazy world out there. Be curious."*

There are no other species on the Earth that come close to the level of curiosity that ours has. Throughout our long tenure on the planet, we've made good use of this trait of ours. It has allowed us to tame fire, find and gather food, defend ourselves from danger, keep ourselves warm in freezing temperatures, keep cool in searing heat, see farther and closer than ever before, to explore the deepest oceans and the farthest reaches of the universe. It has allowed a human to plant a flag on the surface of another world.

The expression of our curiosity has evolved into what we now know as science and engineering. Quite literally, our very survival has depended on our innate curiosity. Without it, we are left as food for predators, lifeless heaps to the elements, or as pawns to those who wish to do ill-will. We have survived as humans, because we wish to learn.

But, can we go through life knowing what a flower looks like, but not knowing why it has the colors it does? Yes. Can we go through life appreciating the beauty of a rainbow, but no adding to this wonder by knowing how those colors form or where they come from? Yes, absolutely. Can we stride through our lives seeing a fabulous red and orange sunset or the shimmering aquamarine of an ocean beach, without knowing the arguably more esthetic beauty of nature's mechanics that actually allow these wonders to occur? Most of us do happily.

So, then, why study these things? Why should the average person investigate their mechanics? There are so many reasons to have a knowledge of physics, but let me mention two. Understanding the way something works does not, in any way, take away from the beauty of something. In fact, just the contrary; *it adds to its beauty and charm*. When we see that gossamer rainbow in the sky, it is a spectacular image, but this is only its surface. What lies behind, is a fantastic interplay between the Sun, its light, water droplets, and the physical interactions between them all. All these ingredients need to come together in such a way to make the recipe for a rainbow. Knowing how they do this adds to our appreciation for the outcome. A beautiful rainbow is still there to be seen in all its glory, but the beauty is now more profound for us all after knowing what lies behind it.

Carl Sagan, the famous astronomer, once said, in a beautifully poetic way:

"The cosmos is within us. We are made of star-stuff. We are a way for the universe to know itself."

That flower you see on the side of the road is also a beautiful object of appreciation, but not just on its surface. There is also an amazing interplay of physics that is occurring that eventually brings to our eyes the purples, reds, blues, whites, greens, and whatever colors the flower may show us. If we study it further, not only do we see and appreciate these, but we will also understand that these colors come to us due to reflections and absorptions of the Sun's light by the plant's fantastic mix of molecules that give it life. Physics gives us these insights.

It's a much more wonderfully expansive world that the science of physics gives us. Without it, we are limited to only a shadow of what the universe has to offer us. With the short time we are given to experience life and universe, why not make the most of it and take advantage of what is on offer.

The other reason I would like to provide for learning about physics was already briefly mentioned. ***It is for our survival***. It is perhaps the only science, outside of biology as well, that has had the most influence on our species' longevity on the planet. It is the science that has provided engineers throughout history with the means to forecast for, build for, and protect us.

There are existential threats that are everywhere around us and, at base, only the principles of the science of physics can help us. Anthropogenic (human-induced) climate change is undoubtedly becoming an increasingly Hydra-headed monster threat to many villages and populations around the world. The physics behind climate science, weather satellites, and meteorological forecasting are essential at predictions and modeling. Solutions also need the science of physics to deal with the climate changes. Trapping and sequestering carbon emissions, or even reducing or stopping them altogether, rely on physics. The changeover to cleaner alternative fuels needs the science of physics to make it happen.

Civilization destroying asteroids and comets exist throughout our own solar system. For all intents and purposes, they are strolling along in our cosmic backyards on very predictable trajectories. Many come very close to the Earth's path and on a regular basis at that. If even one of the larger of these objects were to impact the Earth, the potential destruction to humanity or its civilization is too depressing to contemplate. Yet this scenario is all-too-real and has happened multiple times in Earth's geological history. The most famous event being the asteroid impact that wiped away the dinosaurs and fully 75% of all other species on Earth. However, that extinction event wasn't the only, nor was it the largest, one that has occurred on Earth. With physics, we can find and track these doom's day denizens. We can predict their exact orbits and trajectories and know if, and when, they will make contact with the Earth. This isn't all that physics can provide for us. This science has the power to also help in the *prevention* of these objects from colliding with our planet. Whether the means to do so is to gently push or pull the object into a different route from the Earth's, make a physical impact on its surface to nudge it away, or to destroy it altogether with some sort of explosive or rocket – all of these methods absolutely require physics to achieve.

The universe is, quite literally, screaming out in a commanding voice reminding us that it is a power to be reckoned with. It reminds us that we must use and master its nature to defend ourselves and humanity's survival from that beautiful, but tamable universe that surrounds us.

But let's play devil's advocate for the moment and pretend that humanity isn't as omniscient or as wise as we may think. What would happen if one of these major catastrophes were to really occur? Would physics be able to help our survival then? The answer, even in this case, is yes, but we must act before the event. To ensure greater survival, not only must we manage these existential risks, but we must also find an alternative backup plan for our species. This is to increase our odds of surviving, even if one of these (or another that has not yet been anticipated or known) were to raise its deadly head. The way to do this is to make our species a moving target and not a sitting duck. Multiple colonies spread around the solar system, like on the Moon or on Mars, is the only way. The Soviet rocket scientist Konstantin Tsiolkovsky once said, "Earth is the cradle of humanity, but one cannot live in a cradle forever." Several countries and corporations are currently already investigating and contributing to the effort of colonizing the Moon and Mars. Within a few more decades, humanity will hopefully have its first fully-operational and self-sustaining colony on Mars. Not only will this be yet another triumph for the science of physics, but should be a bright and hopeful sign that our species may survive indefinitely into the future.

The most significant part of all of this, however, is the fact that you, Dear Reader, can be a part of all of it! You should have great pride in knowing that physics is something that endears us all to be a part of one species and that together, we use this tool as both a means for our survival and ends to our curiosity!

Everyone's Field Guide to Physics

Physics is a science that is very diverse. Throughout history, new branches have sprouted and have taken on a life and field of their own. Physics is also a very cumulative science. To be familiar with the area of quantum mechanics, you must first have a knowledge of the main branches of classical mechanics and also some familiarity with electricity and magnetism. Before knowing the details of electricity and magnetism, it would be wise to learn classical mechanics first.

This was essentially how these fields had developed through their respective lines of history as well. The study of motion is what began the true science of physics. The study of motion is also known as classical mechanics. Later, scientists began their inquiries and experiments into optics, electricity, and magnetism. Then thermodynamics followed suit. Finally, in the early 20th century, came the revolutions of Einstein's relativity theories and the mysterious quantum mechanics.

The form that this book adheres to will follow this same progression. However, it will be split up into three volumes. Each volume will treat a different set of topics in physics. We will discuss each of these topics, along with their main ideas and concepts. I've attempted to explain these in clear and concise language. What adds context and further appreciation to these ideas, however, is how scientists came to discover and know them. We will discuss these discoveries and how the experimenters worked to find them. It is one thing to know what inertia is. It is an entirely different thing to know how we discovered its existence.

This volume, *Everyone's Field Guide to Physics, Volume I*, will deal with the physical sciences of classical mechanics and thermodynamics. The reason we have included thermodynamics in this volume, and not later on, as historical chronology would dictate, is due to its similarity with the other subject material within the rest of classical mechanics. So, we will keep the two subjects together in this volume.

Within Volume I, the progression will start with more straightforward concepts first and later deal with the more complicated. Chapter 1 will discuss Newton's laws of motion and similarly relevant topics. Chapter 2 deals with free fall and how objects are affected by it. Chapter 3 will discuss Newton's theory of universal gravitation. A small chapter, chapter 4, will follow and briefly discuss weight and how it is formed. We continue our discussion of motion in a straight line with chapter 5, along with momentum and its conservation law. Rotational movements are treated next in chapter 6, along with its own conservational law. The worlds of work and energy are dealt with in chapter 7. Vibratory and repeatable motions are discussed in chapter 8: Vibration.

In the last four chapters of this volume, we will talk about phenomena in nature that have **emergent properties**. These are properties that do not make their presence known at the individual particle scale, but only appear when there are vast numbers of these particles interacting with one another, like within fluids. So, chapters 9 and 10 deal with fluids and gases. In chapter 11, we discuss temperature, and finally, in chapter 12, we deal with the all-important and universal subjects of heat, entropy, and thermodynamics.

In our next volumes on physics we will deal with later and more recent developments. In Volume II, the discussion will center on Light, Electricity and Magnetism. Essential topics that underpin our current technological and energy economies. Our civilization, quite literally, depends upon these fields for its survival and well-being.

Volume III, our final volume, will discuss the most recent developments in physics and their groundbreaking revolutions. These include Einstein's Relativity Theories (his Special and General theories), Atomic Physics, and Quantum Mechanics.

So, without further adieu, let us begin peeking behind the veil that hides the inner workings of our vast and beautiful universe.

VOLUME I

-

THE FAMILIAR WORLD

CHAPTER 1
THE LAWS OF MOTION

"Nature and nature's laws lay hid in night;

God said, 'Let Newton be' and all was light."

- Alexander Pope

The first stop in our exploration of physics will be with the simplest of the subjects that this science deals with: motion. However, don't be fooled – what we will learn of motion is also some of the most profound, philosophical, and essential of all topics in science. It is the foundation on which all of physics stands. Let's start from our beginnings.

Ancient and Medieval Philosophy of the Universe and Motion

The ideas concerning motion and its causes were never really an issue prior to the time ancient Greek culture and philosophical thought turned its attention to it. Movement was just something that happened and was not significant in and of itself. Yes, the Sun, the Moon, the stars, and the planets all moved around the sky and the Earth on a daily basis, but these realms were not to be questioned with penetrating thought. Instead, they were to be looked upon unquestioningly with praise, respect, and worship. To interrogate into the nature of the houses of the gods may have been thought of as touching on the borders of sacrilege. The gods have their reasons for the motions of and cause for, the heavenly bodies. We have no authority to question these matters and must, instead, be accepting of their providence. Afterall, the Biblical Eve tempted Adam to eat the fruit from the Tree of Knowledge and to consequate the downfall and sinfulness of all of humanity. The best one can do is to make use of their clockwork inevitabilities.

Motions of worldly objects, those more down to earth, may, perhaps, be inquired upon and this is exactly what began to happen in ancient Greek culture on all matters of the world.

Aristotle (384-322 BC), the Greek philosophical student of Plato and teacher to Alexander the Great, whose influence in his time, and even in the later medieval world, cannot be underestimated. Whole societies and eras were to use his teachings as their bedrock understanding of the world. With the exception of the teachings of the Holy Bible, Aristotle's philosophies and writings (and Euclid's mathematics) are what made up the scaffolding and explained the mechanisms that made the universe what it was and how it behaved. (later, the Holy Bible was the supplemental universal manual on how humans were to live within that universe).

Aristotle, after leaving his teacher's Academy, started his own school in Athens called the Lyceum. Here, he had students and teachers discuss, debate, and formulate new ideas about the world. These formulations were collected within his extensive canon of philosophical writings and books. They cover everything from astronomy to zoology, politics to metaphysics, ethics to poetics. Aristotle was the ancient equivalent of the philosophical Renaissance Man.

What were Aristotle's ideas of motion? His concepts of motion exemplify how most academics explained movement up until the time of Galileo Galilei and Isaac Newton. Like most thinkers of his time, Aristotle believed that all the matter of the universe that we experience was made up of four different elements: earth, air, fire, and water. (A fifth element, called "quintessence," also named "aether," was a perfect and unchanging substance and was reserved for the heavens.)

> *"The elementary qualities are four, and any four terms can be combined in six couples. Contraries, however, refuse to be coupled: for it is impossible for the same thing to be hot and cold, or moist and dry. Hence it is evident that the 'couplings' of the elementary qualities will be four: hot with dry and moist with hot, and again cold with dry and cold with moist. And these four couples have attached themselves to the apparently 'simple' bodies (Fire, Air, Water, and Earth) in a manner consonant with theory."* (Aristotle, *De Generatione et Corruptione*, Book II, Chapter 3)

Everything solid, liquid, and gaseous in the world was made up of certain mixtures of these four fundamental parts. Each element had a specific attraction towards either the heavens or the center of the universe and innately moved toward either one if left to their own devices. (The fifth element, quintessence's characteristic directionality of motion for heavenly objects was in perfect circles). So, air and fire had a particular affinity to the sky and, therefore, moved upwards. Whereas, water and earth moved downward, because of their own attraction to the center of the universe. Incidentally, the idea that elements had a kind of built-in "directionality" associated with their natures also very much exemplifies Aristotle's pervasive philosophical tenor that there exists purpose (also known as teleology) in all things. Aristotle said,

"If then, it is agreed that things [in nature] are either the result of coincidence or spontaneity, it follows that they must be for an end…It is plain then that nature is a cause, a cause that operates for a purpose." (*Physica*, Book II, Chapter 8)

So, this is how Aristotle explained the motion of objects that had been left to their own devices. Tip a cup of water over, and the water will, of its own nature, fall downward. Light a fire and the fire will rise. One of fire's purpose was to direct itself upward to the realm of fire.

If an object is at rest, to begin with, it can be made to move by forcing it to do so from another object pushed against it. But here's the clincher in Aristotle's argument of motion: *an object, if it is to continue moving, must be continuously forced to do so.* So, if you have a cup on a table, it will be at rest indefinitely. The only way to make the cup move is to push or pull it, but you must continue to be in contact with the cup if there is to be continued motion. Otherwise, the cup will regain its restful state. A bow's string pushes against an arrow and moves the arrow forward. Now we run into a problem. How does Aristotle explain the motion of the arrow *after* leaving the bow's string? It does, indeed, move after contact with the bow has ceased. When released, the arrow leaves the bow with a certain speed and continues moving. How so, since nothing is touching or pushing the arrow anymore? As Aristotle says,

"If everything that is in motion with the exception of things that move themselves is moved by something else, how is it that some things, e.g. things thrown, continue to be in motion when their movent [contacting mover] is no longer in contact with them?" (*Physica*, Book VIII, Chapter 10)

Despite the tongue-twisting convolution of Aristotle's statement, what he is saying is that this really is a problem that screams out for an answer. Aristotle came up with a contrived solution of his own. According to Aristotle's reasoning, motion can only be given to an object from another by force. The movement, says Aristotle, can only be maintained if the force continues, and the only way a moving object that is not in contact with anything else is kept in motion is from an invisible object or substance that must still be pushing against it. But what substance could this be, and how would it work? Aristotle's answer was air. Once the arrow leaves the string, it continues feeling a force from behind by the push of air. As the arrow moves through the air it pushes aside the air in front, this air then moves along the length of the arrow and makes a kind of U-turn around to the back of the arrow, and subsequently pushes against it. The air essentially makes a circulation around the arrow and maintains its motion. Throw a ball, and it continues forward after leaving the forced contact with the hand in the same way.

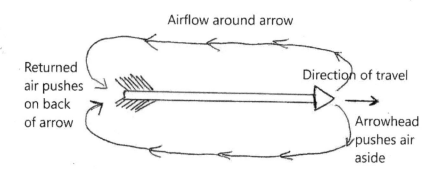

Aristotle's explanation for the motive force of a freely moving object

Aristotle understood that this motion could not continue indefinitely. He saw that the arrow, the ball, or the stone, after traveling some distance, will gradually fall back to the Earth. Aristotle believed the reason for this was the continuously diminishing "motive force" behind the air.

Obviously, this circulation of air hypothesis is a kind of perpetual motion and is impossible in the real world. Not to mention, that the force that the air would need to impart on the object would be much too feeble, anyway. Throw a baseball up into the air and then try, with all your might, to blow the ball forward. How far did you get? Not very far, I would venture to guess. But, Aristotle would lead us to believe that this is precisely what happens when objects move forward in the air.

Thus, this concept stood for centuries after Aristotle's Lyceum teachings. The rise and fall of the Roman empire, and afterward, the passing of the medieval world, swept through history with Aristotle's ideas in tow and still in reverence. His worldview was gripped onto, and brandished, throughout European monastic life and into the medieval universities. **Scholasticism** was the name given to the form of teaching at this time, which reconciled Christian theology with Aristotelian philosophy. Significant figures in this period, whose focus was in scholastic teachings were such influential thinkers as Anselm of Canterbury, Peter Abelard, William of Ockham (of "Ockham's Razor" fame), and St. Thomas Aquinas. For example, Aquinas wrote prolifically, and in his major treatise on Christianity, *Summa Theologica*, he deferentially refers to Aristotle as "The Philosopher." Many of Aquinas' philosophical arguments followed much of Aristotle's reasonings. St Thomas Aquinas and the other scholastic academics made major contributions to philosophical and scientific inquiry that had resonance with academia for many centuries to come.

It is incredible to think that it took all these centuries to question Aristotelian concepts of motion and to admit that they were false. Such is the dampening influences of dogma throughout history. Dogma weighs progress down. Many other examples can be shown where this has occurred, but we will limit our future discussions to only a few.

Experiment

This erroneous thinking was to continue until the brilliant insights of a man who dared to not only *think* about motion, and to ask penetrating questions concerning its true nature, but to additionally *experiment* with it. This man was to change the way we see the world forever afterward. Galileo Galilei (1564-1642) was born and lived in Pisa, Italy and was undoubtedly a true Renaissance Man as he was an inventor, an engineer, a mathematician, a physicist, and an astronomer, and made contributions to all of these fields. Galileo invented special kinds of compasses for the military, a thermoscope[1], he improved upon the refracting telescope and developed parts for pendulum clocks.

Most importantly for our discussion is Galileo's contribution to the scientific method in which he made clear use of mathematics and experimentation to test physical phenomena. He was beginning the sacrilegious investigations into heaven that

---

[1] A thermoscope is a kind of thermometer that uses a fluid that rises or lowers within a tube depending on surrounding temperature. The difference between it and a thermometer is that a thermoscope has no temperature scale associated with it.

the old religious vanguard had implicitly warned against. And these scrutinizing eyes were not very far away.

Galileo worked from a laboratory of his own making, designing and constructing devices and experiments to suit his investigations. Because of these efforts, and unfailing perseverance to find answers and patterns in nature, and to accept them over dogmatic thought, regardless of their possible adversity to current thinking, Galileo is sometimes considered the first true scientist. Indeed, I believe he was. Isaac Newton once said that "if I have seen further, it is by standing on the shoulders of Giants." With his investigations into motion, Galileo himself was one of Newton's "Giants."

Mass

Before we can start talking about the nuts and bolts of our universe and how they are combined to make it work, we must first talk about the nuts and bolts themselves. Preliminaries are a must here. This brings us to our second concept in physics: mass. What is mass?

When we speak of **mass**, we really mean how much matter or stuff is within an object. Mass is physically measured with the help of a balance and is given in units of kilograms (or "kg"). What contains one kilogram of mass? A one-liter bottle of water has about one kilogram of mass. Intuitively, we think of *mass* as synonymous with *weight*, but this isn't exactly correct in the world of science. They are similar concepts, but not at all the same thing. Whereas mass relates to the quantity of matter within the object of concern, weight relates to how hard gravity pulls on that matter. Weight is a force; mass is a quantity.

Take a cup of water and compare it to a gallon of water. Obviously, there is more water in the gallon container and, therefore, more mass than the cup of water. What happens when we have two containers of air, such as within a closed plastic container usually meant for food and another container of the same material, but larger? Does the larger container have more mass of air? Again, intuitively, it turns out that it does.

In those cases, we were comparing apples to apples. Let's compare apples to oranges. How about a cup of water with a cup of peanut butter? Which one has more mass? Well, the peanut butter does, even though we measure equivalent *volumes* of each. There happens to be more matter squashed into the peanut butter volume than there is in the capacity that has the water. This is due to the structures of molecules within each respective container and how the molecules are arranged next to each other. The peanut butter molecules are larger and more densely packed together than the water molecules are. Water molecules are smaller and more spread out within a given volume.

As mentioned earlier, mass is measured with the help of a balance. This is basically a spruced-up version of a simple lever with the fulcrum placed directly under the middle of the lever. Place both masses to be compared on either side of the lever and measure how much deviation the lever makes from the horizontal. The more deviation to one side, the more massive the material on that side is.

Let's go back to the kilogram; the basic unit of measurement for mass. The kilogram is the standard unit in what is called the metric system and is in use by most countries of the world. Sadly, the United States is not included in that grouping and instead uses a system of measurements called the Imperial system after England's widespread use of it. (Fortunately, the UK wisened up and in 1965 adopted the more logical metric system as well). The United States' basic unit of mass is the *pound (lb)*. There are about 2.205 lbs for every kg.

Velocity

Another fundamental concept in science, especially in physics, is that of **velocity**. Again, in colloquial usage, the term *velocity* is synonymous with *speed*, but in physics, they are slightly different terms and need to be used with care. Speed is the distance that is traveled in a certain amount of time. When we say, "the car is traveling at 60 miles per hour", we are mentioning a speed. It travels the distance of 60 miles over one hour. That is its speed. In this case, the car's speed would be written as 60 mi/hr (or 60 mph). However, with speed, you have no idea what direction the vehicle is going. Is it going north at 60 mph, or west, or maybe southwest? We don't know if only speed is given.

This is where velocity comes in. Velocity describes the speed *and directionality* of the thing moving. So, "the car is moving southwest at 60 mph", is a good use of velocity. Speed is how fast something is moving; velocity is how fast something is moving *and* in what direction.

However, one thing that we need to make clear is the use of correct units when speaking of velocity. Again, there are differences in Imperial versus metric units in this case as well. Since velocity is a combination of distance and time, we need to be consistent with the units for each of these concepts before we can begin talking about velocity. In Imperial units (and, therefore, in the United States) the primary distance unit is a *foot* (ft) and the basic time unit is, happily, a *second* (s). The metric base units for these measures is, respectively, the *meter* (m) and the second.

So, in the US, we speak of feet per second (ft/s), but in the world of science, we say meters per second or m/s.

Inertia

Our next stop in our physics preliminaries is a basic, but profound, concept called inertia.

Experiments performed by Galileo established the reality of inertia. **Inertia** is the resistance that an object with mass has to changes in its motion. So, if an object is at rest and not in motion, it will have a certain resistance to move when an attempt is made to do so. Likewise, if an object is moving to begin with, then it will have an innate tendency to resist changing that motion, such as with slowing it down, speeding it up, or changing its direction.

This is a weird notion. Why does an object resist movement? If an object is pushed, why doesn't it just move forward without any resistance at all, even if we remove resistive forces such as friction? It is almost as though objects are somehow bonded or "glued" to space in such a way that prevents their motion and that this bonding needs to be overcome before movement happens. I like to imagine inertia as being a kind of friction against space itself.

On the other hand, if an object is already moving at a constant speed, why is it that it takes any effort at all to slow down and stop the motion? If no force is applied to the object to do so, it will continue moving in a straight line forever. This is a kind of moving inertia. Inertia is strange, indeed.

A thought experiment was described by Galileo that showed moving inertia at work (it isn't clear whether he actually performed this experiment, but it isn't at all inconceivable that he did). He asks us to imagine two ramps (or inclined planes), the left one going down and the right one going up. In between the two at the bottom of both ramps is connected a perfectly flat and horizontal plane (the base).

Place a ball a certain height above the base and on the left downward incline and release it from rest (labeled "1" in the diagram below). Ordinarily, what occurs is that the ball rolls down the left incline at an accelerated speed until it reaches the base level and then begins rolling to the right, but then slows down slightly due to friction between the ball and the base. Once it reaches the right incline it starts rolling up that ramp, but at a decelerating (slowing) speed until it finally, and momentarily, stops ("2"). Where it momentarily stops will be lower in height than when it first started (assuming the right ramp is high and long enough). Then the process will reverse and start again, this time from the right incline. Down the ball accelerates to the straight base slowing down a little, then back up the left slope and slows down until it stops ("3"). Again, the height at this point is still lower than where it started its downward motion from the right incline. This process will continue until the ball eventually stops moving on the base at the bottom of the slopes ("4").

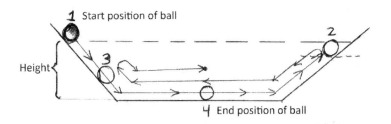

The motions of a ball within a well under the influence of friction.

The draining of the energy and the resultant slowing of the ball to stillness is all a result of the constant friction against the surfaces that the ball experiences as it moves.

Now comes the critical point that Galileo wished to convey. What would happen to the ball if friction was ignored? The answer is that the ball would accelerate down the first incline, roll along the flat base, roll back up the right slope and eventually slow down at the *exact same height as it began with*. Amazingly, the ball would then roll back down this incline and repeat its motion over and over again, non-stop, as shown in label "A" in the figure below.

Still keeping friction at bay, let's imagine decreasing the angle of the right incline and retesting the experiment (labeled "B" in the figure). As the diagram below shows, the ball will continue to roll up the right incline up to the height that the ball initially started at. Now, what would happen if we were to decrease the angle of the right slope to 0 degrees (straight and flat)? Well, since the ball had enough energy to make it up the right incline, when there was a steepness associated with it in previous experiments, it will continue to roll to the right indefinitely until slowed down or stopped by a force (labeled "C"). This is because the ball will never reach the height that it started at and the initial energy that was given it to from the acceleration down the left incline will keep it going forever.

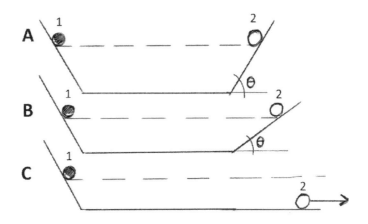

A ball rolling in a well with no friction. The right incline's angle is decreased with each trial.

At the peak of the ball's position on an incline, the force of gravity will budge the ball out of its momentary rest and start the ball rolling down again. So, on the slope, the only thing that gets the ball to change its stillness is a force. Once on the base, the ball will move at a constant speed unhindered. This is the concept of inertia, and Galileo discovered it in his experiments. To reiterate yet again: inertia is an object's resistance to change (either from a resting position or a moving one) unless it is acted on by a force.

Scalars

What is the temperature where you're at? What is your local time? How far away are you from the nearest store? How many books do you have in front of you at the moment? What is the weight of your pet (assuming you have one)? All these questions usually yield one numerical answer. Maybe the temperature outside is currently 68 degrees Fahrenheit. The local time may be 11:00 AM and the distance to the nearest store may be about 2 miles away. I currently have three books on my table in front of me, and my dog weighs about 14 lbs. These single numbers represent a concept in physics called a **scalar**. A scalar is simply a quantity; it is a value that is measurable. Think of a "scale" that you use to measure with, and it will give you an answer as a scalar.

Temperature is a scalar that measures warmth or coolness, and a thermometer is used to make this measurement. Time is a scalar and is measured with a clock. Distance is also a scalar property and is measured in many ways, but a meter stick is a good start. The number of objects there are in an area is a scalar that is simply measured by counting and weight is also a scalar property, measured with a scale. Scalars are all around us.

Forces and Vectors

This brings us to another fundamental concept in physics called **force**. What is a force? In physics, there are several definitions, but intuitively, we can simply think of it as a push or a pull on an object. Gravity is a force and only has the property of a pull between masses. The mass of the Earth has a gravity that pulls you toward its center. When you walk, your shoes push with a certain downward and backward force against the ground. The Earth is pulling you down with its gravity, causing you to have weight. Your weight pushes down against the ground, but the ground is also pushing you up as well (otherwise, you would fall through the ground) and both are considered forces. An airplane stays in the air due to an upward force against its wings called lift. A baseball contributes to home runs by flying through the air from the force of impact of a bat. Forces are pretty easy to identify and to understand.

There is a unique unit associated with a force, and it is aptly called a Newton (N), named after the inimitable and great Isaac Newton. That is the metric unit for force, but in the Imperial system, the pound (lb.) is used, instead. How much force is one Newton? Well, take a medium-sized apple, place it in your hand, and the weight of that apple is, give or take a little, how much force a single Newton is. Newtons and apples go well together!

As you can see, forces are not scalars; they do not just contain a quantity, there is a direction to them as well. Forces are an example of a mathematical concept called a **vector**. They sound scary but are easy to understand. Think of vectors as arrows. They have a certain length (a scalar) and direction. The length of the arrow represents the vector's strength or quantity.

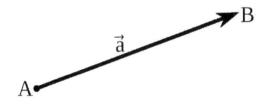

The humble vector with directionality and length

Forces have a specific strength and direction, as well. We've also discussed another vector – velocity.

Mathematically, when we write the symbol for a vector, it is represented either as an arrow above the variable or as the symbol in bold, like so:

$$\vec{F}, \vec{v} \quad \text{or} \quad \mathbf{F}, \mathbf{v}$$

You can manipulate vectors in the same way you do in simple arithmetic. They can be added, subtracted, multiplied, and divided. These are all accomplished in the same way you may imagine. When adding vectors together $(\mathbf{A} + \mathbf{B} = \mathbf{C})$ you place the arrowhead from one $(\mathbf{A})$ against the tail of the other $(\mathbf{B})$ and the resultant vector $(\mathbf{C})$ is the vector that extends from the tail of the first to the head of the second vector (as seen in the diagram below). What if one vector is in the opposite direction as another? You add them the same way: head to tail and the resultant vector starts where the first starts and ends where the second ends. This is essentially the same operation you would perform as vector subtraction. When you take vector A and subtract vector B from it, you are really adding vector A to the reverse (or the flipped version) of vector B: $\mathbf{A} + \mathbf{-B} = \mathbf{A} - \mathbf{B} = \mathbf{D}$.

When multiplying a vector, you are increasing the length of the vector, if the multiple is more than one (for example, 2 X $\mathbf{A} = 2\mathbf{A}$). Lastly, if dividing a vector, its length gets cut proportionately smaller $(eg, \frac{1}{2} \, X \, \mathbf{A} = \frac{1}{2}\mathbf{A})$.

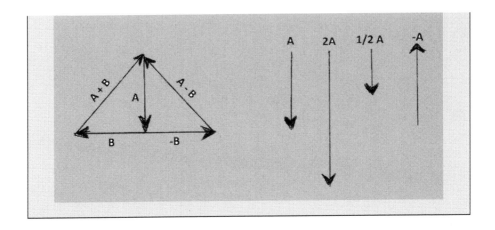

Newton's Laws of Motion

In his celebrated 1728 treatise, *Philosophiæ Naturalis Principia Mathematica*, Isaac Newton stated his three famous **laws of motion**. These laws form the bedrock for all of classical mechanics. The three laws are succinctly stated thus:

1. *Inertial Law*: An object in motion stays in motion, and an object at rest stays at rest unless either one is acted upon by a total force.
   Let's have it in the great man's own words:
   "That is, a particle initially at rest or in uniform motion in the preferential frame continues in that state unless compelled by forces to change it."

2. *Force Law*: An object of mass m, when acted upon by a total force F, will give the mass an acceleration. (Acceleration will be covered in more detail in Chapter 2: Falling Bodies).

And, again, from Newton's translation, which was originally in Latin:

"The alteration of motion is ever proportional to the motive force impress'd; and is made in the direction of the right line in which that force is impress'd."

3. *Action/Reaction Law*: Every force (action) onto another object, has an equal and opposite reactive force.

   Almost the same, word for word, in Newton's tongue:

   "To every action, there is always opposed an equal reaction: or the mutual actions of two bodies upon each other are always equal and directed to contrary parts."

Let us discuss these laws individually and in more detail.

The Inertial Law:

Newton's first law of motion is just a restating of Galileo's inertia. As we discussed earlier, inertia was a concept that did not occur to the ancients, nor to the academics of the middle ages. Inertia is not an entirely intuitive concept and required careful experiment and iconoclastic thinking to deduce it. Usually, we would think that once an object is moved, due to some sort of force pushing or pulling on it, the object will eventually slow down and stop if that same force were to be removed. However, the fact is that there are additional forces that *are* continuously slowing the object down in the real world. These come in such forms as air and water resistance and surface friction. Our everyday experiences all relate to objects immersed in some sort of medium: air, water, and contact with the solid ground. We are not used to an environment where an object moves through a vacuum, devoid of any of these resistances. What happens in that case? Conceptualizing the removal of the surrounding medium through which an object moves, and additionally providing the cognitive leap to imagine the innate character of inertia of an object, required the minds of such thinkers as Newton and Galileo. The object will continue to move indefinitely.

The Force Law:

Newton's second law of motion introduces the nature of acceleration. How are objects (masses) found to accelerate or change direction, if their natural tendency is to stay resting or to continue moving in a steady straight line? The answer is from the application of a **force** to the mass. A force is simply a push or a pull. **Acceleration** is a change in velocity, which means a change in speed and/or a change in direction. When an object has an applied force impinging on it, it will make the object accelerate. If the force pushes or pulls on the object's side, the object changes direction. If pushed from behind (or pulled from the front), the object accelerates, and it speeds up. If the inverse happens, and the object is pulled from behind (or pushed from the front), the object slows down and "decelerates."

The converse of Newton's second law can also be seen: if an object begins to accelerate it is due to some sort of force being applied to it. Slow down, speed up, and/or change direction, and the mass is experiencing a force.

Newton stated this mathematically as[2]:

$$F = m \times a = ma$$

Force (F) is the same as multiplying the mass (m) of the object by the acceleration (a) transferred to it.

The units of force in Imperial units is foot-pounds per second squared (ft-lb/s^2). In metric units (also called **International System of Units**, or **SI** units) this is meter-kilogram per second squared (m-kg/s^2). Luckily for us, there is an individual name for the SI unit of force called, aptly, the Newton (N).

The Action/Reaction Law:

Newton's third law of motion is another bit of counterintuition that only the brain of Newton could discern. Push or pull an object, and that object will push or pull back just as much but in the opposite direction. Push on a wall with your hand. You are applying a force upon the wall, but that wall is, likewise, pushing back on your hand! Sit down on a chair, and your weight (a downward force) impinges on the chair. That chair has an equal and opposite effect pushing back on your

---

[2] When writing letters in an equation they are called variables, and each represents a certain value. So, for example, "m" and "a" represent "mass" and "acceleration", respectively. Now, traditionally and mathematically, when we want to multiply two variables or quantities together, we can either write "m X a", for instance, or we can write "ma". When we place the two letters next to each other we assume that each letter is a different variable associated with a different quantity. When they come together, they are implicitly being multiplied together. It is much more convenient, and less confusing, to place them together, which I will do from this point on throughout the book.

behind in reaction! All these reactive forces are called **normal forces**. Imagine what would happen if this normal force did *not* exist. Your hand would go through the wall, and your body would fall through the chair since there would be no forces to prevent them from doing so! These normal forces are what keep your hand and derriere from moving any farther.

After seeing how these normal reactive forces work, you may say they're reasonable enough. But these are the cases where there is physical contact between two objects applying forces upon each other. However, there are forces between objects where there is no contact whatsoever. These forces are what Newton called "action at a distance," and he was very troubled by them. Newton was so disturbed by this concept that he believed there had to be some sort of invisible and intermediate substance that conducted the force of gravity. These are his words:

> *"It is inconceivable that inanimate Matter should, without the Mediation of something else, which is not material, operate upon, and affect other matter without mutual Contact...That Gravity should be innate, inherent and essential to Matter, so that one body may act upon another at a distance thro' a Vacuum, without the Mediation of anything else, by and through which their Action and Force may be conveyed from one to another, is to me so great an Absurdity that I believe no Man who has in philosophical Matters a competent Faculty of thinking can ever fall into it. Gravity must be caused by an Agent acting constantly according to certain laws; but whether this Agent be material or immaterial, I have left to the Consideration of my readers"* (a letter by Newton to a Mr. Bentley, Feb. 25, 1692/3)

We will return to the concept of action-at-a-distance forces and their transmissions in a later volume. However, other examples of these action-at-a-distance forces are the electric and magnetic forces and gravity. So, how does Newton's third law apply to these forces, where no contact between objects occurs? The same way: equal and opposite reactivity. Imagine pulling on a rope in a game of tug-of-war. The rope pulls back on you. The gravity from the Earth pulls on your body, but your body also pulls back onto the Earth. Your body's force is so insignificant on the mass of the Earth, though, that no perceivable natural reaction takes place, but it's there, nonetheless. Place a magnet near another magnet, and they either feel a push or a pull depending on their respective poles. In both cases, each magnet feels the same force from the other, but in the opposite direction. Magnet A pulls on magnet B, but magnet B also draws on A the same amount. The same thing happens with the electric field and respective electrical charges, which will all be discussed in the next volume.

Centrifugal and Centripetal Forces

We've all seen rides at fairs that spin people around in little 2-person carriages. Many carriages are hanging by cords that when the ride begins, the spinning carriages lift up and the riders feel like they are heavier than they usually are. Then there is the ride, called a "gravitron," that you stand up within and begins to spin so fast that it actually pulls you against its walls. When it is spinning at its fastest, you may actually "crawl" up and down on these walls. This rapid spinning process is even used in medical, and physics, sciences with the use of fast centrifuges. In medicine, blood can be separated into its constituent parts – plasma, platelets, white, and red blood cells – through the use of centrifuges. In physics, enriched uranium from ores, for a bomb or nuclear power plant usage, can be extracted with the use of ultrafast centrifuges.

Another example of the use of forces created from spinning is an interesting one that can be demonstrated at home. Take a small pail of water and attach a rope or cord to its handle. Begin spinning it in a horizontal circle around your body. If it spins fast enough, you may even get it to turn in a vertical circle above your head where the pail is half the time upside down. The neat thing about this demonstration is the fact that if rotated fast enough, no water will fall from the bucket even when it is upside down. How does this happen?

What is being demonstrated are centripetal and centrifugal forces. **Centrifugal force** is that pulling force that is felt against the wall or the seat when spun around. This is the outward force that feels like you will fly away from the center of rotation if you were let go. **Centripetal force**, on the other hand, is that pulling force that is needed to hold on to the cord or the rope when the object is spinning and is in the opposite direction from the centrifugal force. It is directed inward toward the center of rotation.

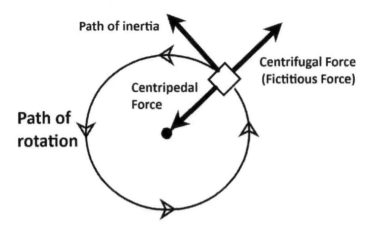

The forces existing and perceived by a rotating object

We say in physics that centrifugal forces are **fictitious forces** (or **pseudo forces**) because it is only an apparent one and is not real. However, centripetal forces are the real thing. Let me explain the difference.

When something is being spun around an axis outside of itself, there needs to be a force directed toward the center of that orbit (a centripetal force) to keep the object rotating. The reason for this is due to Newton's first law of motion: an object moving in a straight line will continue moving that way, unless acted on by an outside force. So, if there was no external force pulling on the object toward the center to rotate it, that object will continue in a straight line instead. This means that a centripetal force is really occurring.

A centrifugal force, on the other hand, is that one felt against the center and outward. This force, however, is only due to the seat or wall being pulled *toward* you! Let's imagine something. You are spinning around in the fair's carriage ride. Imagine now, that all of a sudden, the entire ride disappeared from existence and left you in motion (or, if you like, the cords holding the carriage to the center were suddenly detached). Which direction would you fly away? Would you fly away *outward* in the opposite direction from the center of rotation (labeled "centrifugal force," in the diagram above)? Or would you fly *forward* perpendicular from the line between you and the center (labeled "path of inertia")? The answer: you would fly *forward (ahead)*, not outward. This seems counterintuitive at first sight, but again, this is easy to demonstrate. Test it out to convince yourself. Spin that pail of water, and when you've got it going at a good clip, release the rope. Which direction did it fly away? It will be directed perpendicularly to the initial rope line. Another way of saying this is that it flew away *tangentially* from the circle of rotation.

The competitive Olympic games of discus and hammer throwing use these principles as well. In the future, centrifugal forces will be used to simulate artificial gravity for astronauts in spacecraft. This will be operated the same way as the "gravitron" ride at fairs.

Physics, you gotta love it!

Friction

One of the most pervasive forces that affect all of us surface-dwelling creatures is that posed by friction. **Friction** is a contact force that opposes motion against surfaces. So, as you can imagine, it is a force that can be found everywhere around us. However, I can hear you thinking to yourself, how can friction considered to be a force? Aren't forces something that pushes or pulls? That is correct. Here's the reasoning. Imagine a cup on a table, now give it a push. Friction slows it down to a stop. Now, Newton's first law, his inertial law, states that when something is in motion, it will continue in that motion *unless acted on by an outside force*. Well, that cup does not keep moving, does it? It stops. An external force must have acted upon it. The only action that prevented the movement was the friction against the table. That friction, therefore, is a force. It also follows that the frictional force was *pushing* against the cup to slow it down. Frictional forces are in a special category called "non-conservative forces," which we will go into more detail on in chapter 7.

When two physical objects that are in motion relative to each other are in contact, their surface compositions affect how they will interact. If the surfaces are very smooth, then they are more likely to slide past one another easily. If, on the other hand, the surfaces are rough, they will have a more difficult time rubbing against each other.

Rub your hands together fast, and you'll notice a warming up occurring between the two. This is excess heat that is being generated from friction. Where is the heat coming from? When looking at the surfaces of most objects through a microscope, there is a certain roughness that can be seen. There are bumps and valleys and ledges all over the surface. Imagine placing another surface like this against it. All those jagged and bumpy parts will hook on and lock in and around each other. When the two surfaces try to slide past one another, these areas pull or push and provide that resistance you feel in friction. When they break these connections, a microscopic vibration happens along these rough parts, and this is the source of the heat generated. There are millions, perhaps billions, of jagged areas all vibrating together. When you initially pushed the object against the floor, there was a motion you gave to the object, and therefore, a certain amount of energy transferred to that object. All that motional energy gets transferred again to the heat in the friction. Because there is a continuous drain in the motional energy to resistance, the object slows down and gets wholly drained of motion.

Every combination of two materials has a value associated with them that represents their rubbing resistance. It contributes to their friction and is called their **coefficient of friction**. The value is usually between 0 and 1. Zero represents no friction at all and 1 is when the frictional force is the same as the force of the weight of one of the objects on the other surface. There *can* be coefficients of friction that are higher than 1, such as when silicone rubber rubs against another surface of rubber. Those two kinds of surfaces will, indeed, be tough to rub against each other.

There is an extraordinary state of matter called **superfluidity**, where absolutely no viscosity occurs (see chapter 9). Liquid helium is in a state of superfluidity, and it is in this state that matter exhibits zero friction.

CHAPTER 2
FALLING BODIES

"...and yet it moves."

- Galileo Galilei

Free Fall

Let's imagine standing on the outside of the eighth (top) floor of the bell tower of the Leaning Tower of Pisa. We hold in one hand a bowling ball and in the other a feather. Hold them over the ledge (challenging to do in practice but thank goodness it's a thought experiment!). Release both at the same time to allow them to fall - to **free fall**. What do you suppose happens to each? Which one hits the ground first? My guess is that you said the bowling ball. Correct!

Next experiment: same place, but bowling ball in one hand and a balloon blown up to the same volume as the bowling ball in the other hand. Release. Which hits bottom first? Correct again!

Third experiment: same place, bowling ball in one hand, but a plastic ball the same size and shape as the bowling ball. The plastic ball is filled with water but is lighter than the bowling ball. Your bowling ball arm is tiring and shaking. Release them. Which one lands first? That's okay; take your time thinking about it. In the meantime...

Fourth and final experiment: after the exhausting walk up the spiral staircase to the eighth floor of Pisa's vertically-challenged tourist attraction, and a thoroughly sore bowling arm, you again reach out from the banister ledge and hold out the bowling ball and a feather in the other hand. Hold your breath, because now all the air in the atmosphere has been sucked away and you are now in a vacuum.[3] Release the objects! Which object hits the ground first this time?

Okay, so the story of Galileo ascending the Tower of Pisa and performing a similar experiment (with the atmosphere included) was – or would have been - an important one to attempt but is probably apocryphal. While we pretty much knew that both the first and second thought experiments previously mentioned would end with the dreadful bowling ball impacting the tower ground first, it isn't as clear that the third one will yield the same results. While the feather and the balloon continue blowing around in the air, have you ventured a guess as to the solution to the third experiment? In reality, the bowling ball will still make the first crater, but just barely.

All these examples are experiments in "free fall," where the objects dropped are under the free influence of gravity. Let's start from the beginning again. Not in Pisa, but in Athens, with Mr. Aristotle. His intuitions and teachings reflect that when given any two objects, the heavier one will fall faster. His reason? Lighter objects, he said, have an innate tendency to fall slower due to more of the lighter materials within it: air and fire. Both of these materials, as mentioned earlier, were thought of to have an affinity for the sky, whereas the heavier elements of water and earth had an affinity for the center of the Earth. If an object has

---

[3] Yes, I know: holding your breath is not the only thing you will need to do under these circumstances. I didn't mention the horrific facts that your blood will boil away into the vacuum, not to mention all the other liquids contained in your body, and that terrible things may also happen to your eyeballs (and therefore your vision). But did I forget to mention that this was a *thought experiment*?

more earth or water mixed in within it, then it will tend to be heavier and thereby fall faster to the Earth.

What happens in the fourth thought experiment above, where we imagined dropping a bowling ball and a feather within a vacuum? In Aristotle's view, it would not matter if the air were to be removed from the surroundings, because the bowling ball will still fall faster and hit the ground first due to its higher weight.[4] The feather will fall slow and gently and hit the ground last even in a vacuum. Was Aristotle correct?

Inclined Planes

Well, let's time travel about two thousand years forward to Pisa and into Galileo's mind, where we will soon get our answer. Galileo disagreed with Aristotle's reasoning for free fall. Aristotle believed that objects that were heavier fell faster, whereas Galileo surmised they both dropped at the same rate, regardless of weight *if the air was removed from the environment of the falling bodies*. This was a crucial point, as air resistance, according to Galileo, is the factor that influenced the observable rate of drop for objects, not their weights.

To test this experimentally, it is said that Galileo dropped two differently weighted, but still heavy, spheres from the top of the Tower of Pisa and recorded their impact with the ground and found that they both hit at the same time. He did this in such a

---

[4] This isn't entirely true, because Aristotle believed that a void (vacuum) could not even exist in nature in the first place. He believed "nature abhorred a vacuum" and if a void were to begin to be formed, matter very promptly will fill in to replace it. We can just say that to a first approximation, Aristotle believed that air had nothing to do with the speed at which objects dropped, unless they were themselves composed of air.

way, the story goes that he removed the effect of air drag from the experiment by using the more massive weighted balls. As mentioned, this story may never have happened in reality. However, Galileo did experiment with rates of fall of objects in his own laboratory.

As you know, when dropping an object to allow free fall, the object falls very quickly. So quickly, in fact, that it poses problems with timing the fall and gauging the rate. In Galileo's time, there was no such thing as a stopwatch or a timer that was very accurate. He had to get creative with his timing.

First problem: was there a way to slow down the rate of fall of a free-falling object? Yes, there is, and Galileo found the way to do it: with inclined planes. An **inclined plane** is simply a ramp. Place a ball on the inclined plane and allow it to roll down starting from rest at the top and it will accelerate down the incline.

Second problem: was there a reasonable way to track the time that the ball rolls down the ramp? Again, yes there is, and Galileo took advantage of a device called a clepsydra, also known as a water clock. The clepsydra is an ancient time-measuring device and measures water height within a collecting bowl as it progressively gets filled from another container that slowly releases its fill of water. It's a bit like an hourglass for water.

Now that he had both parts of the experiment ready to go, Galileo began testing and measuring. His idea was to place a ball at rest on the top of the ramp, release it, then start timing the "fall" until the ball reached the end of the ramp. He repeated this experiment many times and with differing degrees of ramp angulation (steepness). If you can picture a very long ramp, it would take a longer time for the ball to completely roll along its entire length. If you keep the height of the ramp the same but shorten the ramp's incline, then its angulation gets steeper. The steeper the slope becomes, the closer it gets to 90 degrees from the horizontal table. As you approach 90 degrees, you also get closer and closer to approximating real free fall.

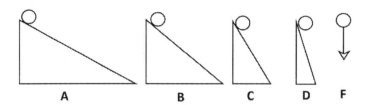

Ball drop experiments on different angled inclined planes (A-D). The free fall of a ball is represented in F.

Acceleration

Galileo found a surprising result with these experiments. What he found was that a pattern emerged that did not at all depend on the steepness of the ramp. That same pattern will appear if the ball was dropped and allowed actual free fall. Let's analyze this pattern to appreciate it better.

Galileo made sure that he was able to mark off a few equally spaced time units for the entire time that the ball rolled along the ramp. For example, a few second units or a few half-second units. This was to ensure that there would be enough data points to measure along the way. Next, he made sure there was a way to measure the length along the ramp with accuracy, like equally spaced markings along the slope: 1 cm, 2 cm, 3 cm, etc. Then he released the ball from rest and started timing the ball. After every equal measure of time (½ second, 1 second, 1 ½

52

seconds, etc.) he made a note of the distance the ball traveled along the ramp. What did Galileo find in his testing?

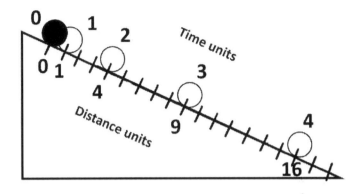

The relationships between the time and distance that an accelerating ball rolls down an inclined plane.

After one unit of time had passed, one unit of distance was traveled. After the next unit of time had passed, an *additional three* units of length were crossed. The following (or third) unit of time passed, and *five more* units of distance were traveled by the ball and so on. The pattern that Galileo discovered was that the distance that the ball had increased, after every unit of time that had passed, had amounted to an odd sequence of numbers: 1, 3, 5, 7, ...

This table is an idealized example of what Galileo had found:

Time Units (from release of ball)	Distance Units (from starting point)
1	1
2	$1 + 3 = 4$
3	$1 + 3 + 5 = 9$
4	$1 + 3 + 5 + 7 = 16$
5	$1 + 3 + 5 + 7 + 9 = 25$

Notice any pattern here? A surprising pattern that is found within this data is the fact that the *total accumulated distance* traveled progresses at an exponentially increasing amount: 1, 4, 9, 16, 25,... If 1 second has passed, then the ball travels $1^2 = 1$ unit distance. If 2 seconds have passed, then $2^2 = 4$ units of length have been crossed. If 3 seconds have passed since releasing the ball, it travels $3^2 = 9$ units of distance along the ramp, and so on. Another fantastic fact: all of this data still holds true even when the ramp's incline changes; it makes no difference what angle the ball rolls. Can you guess the other characteristic that made no difference in Galileo's ramp experiments? Wait for it:...that's right, *weight*! So, let's get rid of the ramp altogether and you get free fall.

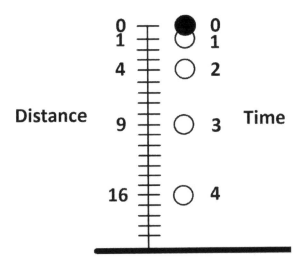

Distance

0
1
4
9
16

Time

0
1
2
3
4

Free fall of a ball dropped from a height.

Free fall is an accelerating phenomenon, and if done in a vacuum on Earth, or on Mars, or in the Andromeda galaxy, two objects next to each other will fall at the same rate of acceleration, regardless of their respective weights. One can be a bowling ball, the other a feather; they will fall together. "Birds of a feather…"

Here's an application of this bit of mathematical esoterica: if you want to impress your friends and family at your next party, take the nearest cocktail napkin and scribble:

$$d \propto t^2$$

Mention that if you were on an airless Earth and you were to drop that napkin, then the distance "d" it would fall is proportional to

("$\propto$") the square of the time it was falling. Tell them that some guy by the name of "Galileo" told you so.

Free fall is an accelerating phenomenon. **Acceleration,** then, is a continuously changing velocity. If the velocity changes speed or direction, it is accelerating. And remember that velocity is a vector? Well, a changing vector is still a vector. So, we can add acceleration to our list of vectors as well.

Vectors: $\overrightarrow{Velocity},$ $\overrightarrow{Acceleration},$ $\overrightarrow{Forces}$

In this chapter, we have discussed the differences between what the ancient and medieval world thought about free fall and compared it to the more modern experiments performed and demonstrated by Galileo. We found that free fall was an accelerating phenomenon and that it does not depend on the weight of the object falling. This brings us to our next question. What produces that free fall in the first place? We will gravitate to that answer in the next chapter.

CHAPTER 3
GRAVITATION

"If I have seen further it is by standing on the shoulders of Giants"

- Isaac Newton

Ancient Tendencies

Okay, so why do things fall down? We all know the answer, and this brings us to the next concept to explore - **gravitation**. We all experience it but take little notice of its effects. It surrounds us and flows through us. It is the invisible pull that is felt on all matter from all other matter, but mainly, at least in our local environment, from the Earth itself. It's that heaviness you feel when attempting a push-up, a pull-up, lifting an object, or simply walking up stairs.

However, as we'll see later, gravitation turns out to be one of the most mysterious phenomena in the universe. Yes, it has been described very well and with exceedingly accurate precision. And, yes, its effects have as well. It is the nature of gravity that is most mysterious about it, but for now, we will leave that aspect of it alone and, instead, focus on gravity's more relatable effects.

The ancients had their own ideas on what constituted the reasons for material objects to drop when released (as discussed in the previous chapter). Aristotle is our go-to guy for our representative "ancient." Due to his massive corpus of writings on very diverse subjects in the realms of science and philosophy,

I believe this is a wise decision. This is not to say that Aristotle's ideas on all matters of physics, astronomy, and philosophy were consistent and equivalent to those of his contemporaries. Indeed, they were not. However, his ideas happened to be the most influential throughout much of history and were the concepts that won the day. They were carried on in most higher institutes of learning throughout most of the Western world for literally millennia. Not to mention that, again, Aristotle's extant writings are very extensive and more complete than most others of his time.

Aristotle elucidated in *Physica*, his work on motion and material objects, that there are layers of the universe where each element's resting place belongs. Aristotle did believe that the Earth was spherical, and its center coincided with the center of the universe. This was the location where the element of earth tended to move toward. This was the resting place of Earth and was the reason that the entire Earth does not move away from it. Aristotle did not elucidate why the element of earth happened to be attracted to the center of the universe. Surrounding the Earth, in the next layer, is the spherical shell where water tends to rest and is where water likes to flow or drop to. Water rests on the surface of the Earth and does not drop below this. Above this layer is the next resting sphere and incorporates air. Blow bubbles of air underwater and where do the bubbles go? Upward, toward the air above. Outside and surrounding this layer is the layer of fire. When a fire is burning, it points up and filaments of fire, when broken away from the main body, tend to move upward through the air. Again, this showed the tendency of fire to move above the air. Lastly, is the resting place of the heavens and the sky, where an element called quintessence resides. The motion and nature of quintessence were not like that of earth, water, air, or fire. It was a substance that made up the heavenly bodies and deserved a much more spiritual and divine emperium.

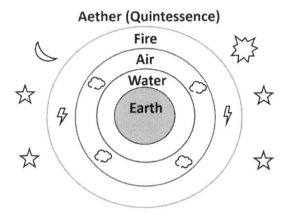

The four pure elements of Aristotle and where their layers reside.

Mixed materials containing mostly earth or water will drop downward, whereas those that mostly contain air or fire will do the opposite and rise skyward. (Quintessence is never experienced as it is always above the fire realm and was the material of the gods, stars, and planets). The heaviness of an object was attributed to the densities and ratios of each of these elements within a particular material, and their natural tendencies to move toward their resting places in the universe.

Why did the objects of the sky (stars, planets, Moon, Sun) not fall down to Earth? As mentioned, these objects were made up of a different sort of element called quintessence whose resting place was above the layer attributed to fire. Also, quintessence had another peculiar characteristic: objects composed of it did not move in straight lines, like up or down, but instead moved in circles around the center of the universe. This explained why heavenly objects moved in such a way around the Earth and did not fall to the center. Quintessence was an eternally existing substance and was perfect (what else could be expected of the material of heaven and the gods?). The most perfect geometrical object, the ancients thought, was the circle and that of a solid was a sphere. Both of these perfect figures happened to reside in the sky. The spherical Moon and Sun; the orbits of the heavenly objects around the Earth. These objects were perfections personified.

'And yet it moves"

So, that was how the ancients touched upon and explained gravity in their time. Believe it or not, this is how it essentially remained for another two thousand years! What had changed that would have unhinged such deeply ingrained and long-lasting notions of the universe? The answer was experimentation and a bold approach to critical inquiry. In other words, modern scientific methods. But I say bold because it was about two thousand years after Aristotle's time when the Catholic Church held sway, and its dogmatic influence was felt by the entire western world. Anyone stepping out of line from the Holy See's world view could expect to be promptly forced back into the episcopal dominion. If they refused to re-enter the Church's universe, they were decidedly taken out of this one. These threats were real for everyone, and Galileo was no exception.

Because of the views he expressed in his writings, especially those on a sun-centered solar system called **heliocentrism** (also called the Copernican system, named after Nicolas Copernicus, who developed the idea and published it in 1543 several weeks before he died), Galileo caught the attention of the Roman Inquisition. How dare he remove the central importance of Earth from its Biblically-proclaimed immobility as the center of the universe! In his day, the two models of the solar system that most academics supported were the Earth-centered **geocentric** system and the hybrid Sun- and Earth-centered **Tychonic** system, named after the Danish astronomer, Tycho Brahe, who had first proposed it 1588. (The complicated Tychonic system also held that Earth was the center of the solar system as well, but that the Sun rotated around Earth and all other planets orbited the Sun).

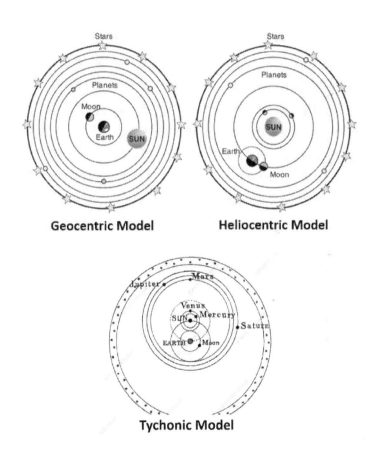

Geocentric Model   Heliocentric Model

Tychonic Model

In his 1632 book *Dialogue Concerning the Two Chief World Systems*, Galileo argued for the Copernican system. However, Pope Urban VIII became very much opposed to it. Galileo was subsequently ordered to stand in front of the Inquisition on 1633. He did so, and throughout the interrogation, he defended his innocence and claimed to never have intentionally sided with the heliocentric perspective. On June 22, 1633, he was read his sentencing and the Inquisition's findings. This included the accusation that Galileo was "vehemently suspect of heresy" and that he, therefore, must "abjure, curse and detest" his earlier support of heliocentrism. A day later, his sentencing was downgraded, and he was required thereafter to go on house arrest for the rest of his life.

It has been said, and again this is another one of those apocryphal scenes that wish to embellish an already dramatic story, that when Galileo had to recant on his views that the Earth does, indeed, move, he mentioned under his breath, "And yet it moves." The 17th-century portrait of Galileo Galilei, possibly by the Spanish painter Bartolomé Esteban Murillo, pictures the scientist in his prison cell with the words "And yet it moves" scratched on the wall next to him.

Despite his views on cosmology and the movements of the celestial bodies, he did not make a connection between them and falling bodies on Earth. Concerning gravity, Galileo established the equivalence of the rate of falling bodies regardless of their weights. These experiments were discussed in detail in chapter 2. It would take a different mind to force the connection between sky and Earth.

Newtonian Gravity

The next monumental jump in our understanding of gravitation came from Isaac Newton. In his 1687 magnum opus, *Philosophiæ Naturalis Principia Mathematica* (Latin for

"Mathematical Principles of Natural Philosophy"[5]) or "the Principia," for short, Newton laid out his three famous laws of motion, his theory of gravitation, and other mathematical treatments in physics and astronomy. Some of the Principia's subjects deal with a more detailed explanation for the motions of the Moon and planets and the cause of the ocean tides on Earth. Newton's intellect was so great that not only was his science of physics and astronomy the base from which all later physical sciences lay, but he also developed the field of calculus and opened up the world of changes to mathematical treatment and investigation. (Newton's competition was in the continental polymath and philosopher Gottfried Leibnitz (1646-1716) in Germany, who also developed the field of calculus at the same time. There was fierce debate between the two as to who was the first to discover this math.)

Isaac Newton was himself a polymath and dabbled in alchemy and Biblical interpretation. He was obsessed with trying to find hidden messages and prophesies in the Bible and even believed that the end of the world in, one form or another, was due to occur in the year 2060 AD. A graduate of Trinity College, Cambridge University, Newton had said that "if I have seen further it is by standing on the shoulders of Giants." He was himself, a Giant that others afterward have relied on to progress science further.

We had discussed Newton's laws of motion in chapter 1. Now we will focus on his theory of gravitation. Everyone knows about the story of how Newton was sitting next to an apple tree meditating on his thoughts when, suddenly, he noticed an apple fall from it (or did the apple hit him on the head?). This event

---

[5] In Newton's time, and earlier, the fields of science were generally referred to as "Natural Philosophy", as they had originally evolved from philosophical investigations. It was only later that the descriptor "philosophy" was replaced by "science".

caused Newton to cogitate on the nature of gravity and asked if the force that caused the fall of the apple had anything to do with keeping the Moon in its orbit around the Earth. Surprisingly, this story is one that, indeed, actually happened and is not at all apocryphal. Newton himself had related the story.[6]

Newton derived his law of universal gravitation from empirical observations. There are several key features of this law. Let's briefly review them and then we'll discuss each in more detail.

1. **Attractive**: Newton said that gravity was an attractive force that existed in all matter. The mass that matter contains attracts all other masses in every direction around the object.

2. **Universal**: Gravity is a force felt across the entire universe. So, the gravity that your body is producing travels across the universe. Yes, you read that right – across the entire universe.

---

[6] Though, in his reiterations, no apple "hit" him on the head! The following quote is from one of Newton's acquaintances, one William Stukeley, who, during a conversation with Newton in Kensington on April 15, 1726, related thus (retaining his English verbiage):

"we went into the garden, & drank thea under the shade of some appletrees, only he, & myself. amidst other discourse, he told me, he was just in the same situation, as when formerly, the notion of gravitation came into his mind. "why should that apple always descend perpendicularly to the ground," thought he to him self: occasion'd by the fall of an apple, as he sat in a comtemplative mood: "why should it not go sideways, or upwards? but constantly to the earths centre? assuredly, the reason is, that the earth draws it. there must be a drawing power in matter. & the sum of the drawing power in the matter of the earth must be in the earths center, not in any side of the earth. therefore dos this apple fall perpendicularly, or toward the center. if matter thus draws matter; it must be in proportion of its quantity. therefore the apple draws the earth, as well as the earth draws the apple."

3. **Reciprocal**: Any mass that you choose has an attractive force upon any other mass in the universe and, reciprocally, those different masses have an opposite attractive force upon the first.

4. **Directly proportional to the masses multiplied together**: The more significant each of the masses is, the higher the forces they each experience from one another.

5. **Indirectly proportional to the square of the distance**: The larger the distance the exponentially smaller the force.

6. **Force of gravity involves a "gravitational constant"**: A constant is simply a number that is always multiplied to another to get its value "started" at some base value. Gravity has one of these constants, and its value had to be determined through meticulous experimentation. However, instead of this constant being large, it is an extremely, a mind-bogglingly, small number (see footnote on page 44).

Each of these concepts starts out easy but get progressively more challenging to understand as we move from key concept 1 to 5. However, we'll attempt to make some headway of them. Afterward, you should have a good grasp of Newton's notions of gravity.

Gravity is attractive.

This is the reason for "what goes up must come down." This is the most intuitive of the 5 key characteristics of gravity. Newton just took it for granted that mass innately produces an attractive force. How it was done, he had no idea. (The world would have to wait for Albert Einstein for that explanation). Gravity does not have a plus and a minus sign attached to it; just a minus sign.[7] If gravity were anti-attractive (or "repulsive"), then we would never expect planets to form or stars to shine, both would explode outward, and stars would no longer have the gravitational energy needed to burn their fuels to shine. Nor would the galaxies ever hold together. The universe would end up being a pretty dark place, indeed. At least we wouldn't have to worry about the voracious activities of black holes or the incinerating possibilities of supernova explosions. But then again, if gravity were a repulsive phenomenon, we wouldn't be here in the first place to admire that darkness.

Gravity is universal.

Thus, Newton's "law of *universal* gravitation." Some of the "giants" that Newton said he stood on, and saw further because of, were Copernicus, Kepler, and Galileo. Each of these scientific greats made significant contributions to our understanding of astronomy and more specifically, how the planets, Moon, and Sun moved amongst themselves. They were all heliocentrists and knew that the Earth was only one of the planets that orbited around the Sun and that the Moon orbited around the Earth. All the heliocentrists knew that it was physical laws that kept the worlds orbiting the central sun. This was contrary to the previously held belief that the angels of heaven kept the planets moving with the beating of their wings behind them.

---

[7] Interestingly, there is a possible force in nature that has recently been discovered that acts like an antigravitational force. This is called **Dark Energy** and it permeates the entire universe. However, it is a force that emanates not from mass, like gravity, but instead from the fabric of space itself. Whatever dark energy is, it is causing the universe to expand at an accelerating rate. Its nature is very mysterious and is one of astrophysics' greatest conundrums. Multiple astrophysics research projects around the world are actively studying this fascinating mystery.

Copernicus developed the heliocentric model or the sun-centered system. Galileo discovered that Jupiter had moons of its own that orbited that planet in what appeared to be a mini solar system of its own. **Kepler** refined the Copernican model further and developed three laws of planetary motion of his own. (Kepler's three laws of planetary motion are explained in detail in Appendix A).

Newton saw that there was a unifying way to explain the fact that all the planets orbited a central more massive object (the Sun), that the Moon revolved around a central larger object (the Earth), and that objects fall toward the Earth (a much larger object than the one falling). Newton reasoned that if the same force that pulled the apple toward the Earth was the same one that kept the Moon orbiting, instead of flying away into deep space, then it seemed that this force had quite a distance of influence. This was even more evident when considering the attractive force that held the planets in their orbits around the much more distant Sun. Gravity, it appeared, was a universal phenomenon.

Gravity has a reciprocal influence.

Derived from one of his three laws of motion, gravity's reciprocity comes from Newton's third law of motion that states that every action has an equal, but opposite reaction. The action, in this case, is the force of gravity. When talking about two massive objects, A and B, object A's pull on B can be considered the action. The reaction happens in the opposite direction and is the pull that is felt on A from the gravity coming from B. The amount of force that is felt on both A and B is exactly the same but in opposite directions.

The amount of force felt is directly proportional to the object's masses multiplied together.

Now comes a little bit of mathematics to help explain this aspect of gravity. When speaking of two objects, again A and B, with the gravity from each pulling on the other, we eventually need to know how much attractive force there is. Experimentally and intuitively it happens to be dependent on how much matter each has within them. In other words, how much mass. The larger the mass that A and/or B has, the larger the force that is felt between them (this is also assuming we are keeping both masses at the same distance from each other). This change in size and the change in gravity's force is also proportional, which simply means that if you were to double the size of the mass of one object, its attractive force also doubles in strength. Halve the size and halve the strength. Triple the size, triple the strength, etc. This is what is meant by directly proportional. When you increase one variable, another variable increases by the same factor.

Mathematically, you can minimize this expression symbolically as[8]:

$$F_g \sim m_A \cdot m_B$$

In words, this means "the force of gravity *is directly proportional to* the mass of object A times the mass of object B." We say, "proportional" ($\sim$) as opposed to "equal" ($=$) because there is more to figuring out the actual value of the gravitational

---

[8] The little dot ("$\cdot$") between the m_A and the m_B represents multiply and can be convenient to use, and clearer to see, when placing two potentially confusing variables together.

69

force than just knowing the masses of the objects involved. One other piece of the puzzle is the distance between the objects.

The amount of force felt is indirectly proportional to the square of the distance.

When it comes to figuring out gravity, the masses of the objects aren't the only thing that matters! Take object A and object B and separate them farther and farther away. What happens to the force of gravity between them? It decreases. So, increasing distance decreases the strength of the force. This relationship is called "indirect proportionality." However, there's something more: when increasing the distance by twice as much, the strength doesn't decrease by half as you may expect, but by a *fourth*! Increase the distance by three times farther away, and the strength of the gravitational force decreases by a *ninth*. When bringing the masses closer together, the opposite happens. Decrease the distance by one half, and the strength increases by four times. Decrease distance to one third, and the intensity increases by nine times. Instead of a *direct* proportionality, there is an *inverse* proportionality. When one variable increases, another variable decreases.

So, what this means mathematically is that the strength of the force of gravity is "inversely proportional to the *square* of the distance" separating the objects. In symbols, this is written:

$$F_g \sim \frac{1}{d_{AB}^2}$$

The gravitational force (F_g) is inversely proportional to the square of the distance separating mass A and B (d_{AB}^2).

This is called an **inverse square law,** and many phenomena in nature follow it. The reason it works is simply because the farther away from the source you are (the gravity that comes from a mass, in this case), the more spread, or thinned, out in space the source's force is. Light follows this law as does sound and electric and magnetic fields (all of which will be covered in the next volume). The cause for the law displaying a "squared" phenomenon can be seen by the figure below:

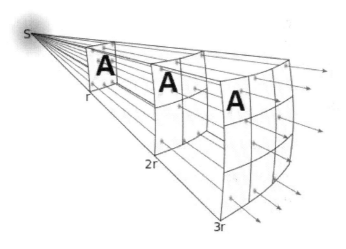

Inverse Square Law. Each distance radius "r" away from source S provides fewer lines of force to pass through each unit of area A.

The object at "S" is the source, and each "r" represents different unit distances from the source. Each square "A" represents the same unit area size. Each arrow in the diagram represents one unit of gravitational strength. The more bunched together the arrows are within a single square area, the stronger the source is (gravity, in our case). As you can see, at distance r, *all* the arrows pass through A; gravity is stronger there because more arrows are passing through a single unit of area. Move out to 2r, and you can see that only ¼ of the total arrows pass through each A. Move all the way out to 3r and each area A now has about 1/9 the number of arrows (or gravitational strength) passing through. This is the physical reason why inverse square laws exist.

The Gravitational constant, G

The last piece of the gravitational force puzzle is the constant that is associated with it: G, the **gravitational constant**. We will discuss this constant in the next section.

So, let's attach all the pieces together to compose our final form of Newton's theory of universal gravitation. Here it is, in all its grandeur and magnificence:

$$F_g = G\frac{m_A \cdot m_B}{d_{AB}^2}$$

Remember, that since F_g is a force, it is measured in Newtons.

Later developments by Albert Einstein improved upon Newton's concepts of gravity, and we will discuss these newer ideas in volume 3 of this series.

The Cavendish Experiment

The British scientist **Henry Cavendish** (1731-1810) designed and performed the very first experiment to test Newton's law of gravitation. Its purpose was to attempt to measure the strength of the gravitational force between two lead balls. The experiment was performed in 1797 by an elaborate setup using very precise measurements. This precision was required because it turns out that gravity is an extremely feeble force. It may not seem so since we are all pulled down to Earth in a very obvious way. But the Earth is very large and has a lot of mass to gravitate us toward it. Place a bowling ball on the ground close to the foot of a mountain range. Is there any noticeable movement or attraction of the ball to the nearby mountain range? I know this example may seem absurd, but there is certainly gravity between the two masses. This just demonstrates how meager that force really is.

Cavendish used what is called a torsion balance for his experiment. The way his torsion balance worked was to suspend a six-foot horizontal wooden rod above the ground using a thin wire attached to its very center. Suspended from each end of the wooden rod were small lead balls. The balance was allowed to twist around the wire ever so subtly. Outside of this setup, were two other, but larger, lead balls fixed, immobile, and close to the smaller lead balls.

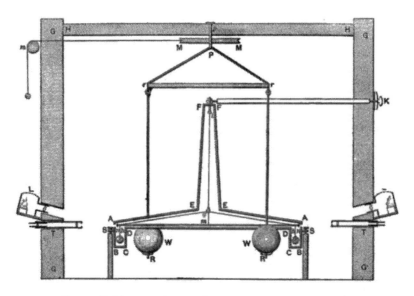

Cavendish's experimental apparatus to measure the force of gravity

Cavendish made sure there were no influences from static electricity, humidity, temperature differences, etc. Cavendish knew ahead of time the torsional strength of the suspending wire, which depended on how much the wire twisted; the more twisting, the stronger the force required to turn it. He also knew the masses of each of the lead balls and the distances between each of them. Sure enough, Cavendish saw the gravity of the lead balls at work when the apparatus began to twist until the twisting (torsional) strength of the wire was equal to the force of gravity between the balls.

From Cavendish's experiment and others, one can accurately determine the value for the gravitational constant (G).[9] Indeed, the gravitational constant is extremely weak. In fact, we know of four fundamental universal forces: gravitation,

---

[9] The current measured estimate for G is 6.674 X 10^{-11} m³/kg s (from the 2014 Committee on Data for Scientific and Technology - CODATA).

electromagnetism, the weak, and the strong forces. Of these four, gravity is by far the weakest of them all. We will come to each of these forces in due time.

For now, let's have a quick, simple, but mind-blowing, demonstration of the weakness of gravity. Take two small magnets. Hold one in your hand and keep the other on the ground. Slowly drop the one in your hand closer and closer to the other. Eventually, you will get to a certain distance of separation between the two magnets where they will attract one another and stick together. Just imagine that. It took the force of a simple, small magnet to pull up on the magnet on the ground and to overcome the strength of the entire planet Earth's gravity pulling it down! Think about that for a bit. Gravity is a ghostly and feeble thing, indeed.

But it is due to gravity that we owe our weight, and this is where we will turn our attention to next.

CHAPTER 4
WEIGHT

"Gravity cannot be held responsible for two people falling in love."

- Albert Einstein

What is Weight?

When we talk about **weight**, we mean how heavy an object is. However, weight is a malleable phenomenon; it is something that can change. For example, a person on Earth may weigh 150 lbs., but will he weight the same on the Moon? No, his weight will then be about 25 lbs. That's how we were able to get those iconic videos of astronauts bunny hopping on the Moon. In fact, a person even weighs slightly less on a mountain peak (or upon an airplane) than at the surface of the Earth. It wouldn't be much of a difference, but it still exists. If that 150 lbs. person (at sea level) was to climb up to an altitude of 10,000 ft they would then weigh 149.92 lbs.

Why is this? It is because weight comes from the force of gravity on the object. The stronger the gravity, the greater the weight the object has. Every object has a certain amount of matter within it. This is called its mass. The mass of an object never changes and will remain the same whether the object is on the Earth, on the Moon, or in the Andromeda galaxy. Remember that *distance* from a gravitational source (like the Earth) also makes a difference in the strength of the force of gravity. So, 10,000 feet away from the surface of the Earth will have a slightly lower gravitational force in the vicinity.

Also, the force of gravity changes depending on the *size* of the objects. The gravity between the Moon and the person is 1/6 the gravity between the Earth and the person. The mass of our Moon is 1/6 that of the mass of the Earth.

The way to calculate the weight of an object is by multiplying the mass of the object by the gravity in the area (called the **"gravitational acceleration,"** g for short):

$$W = m\,g$$

The variable "g" represents the value of acceleration that a body would experience if in free fall in the area you are located at due to the gravity.[10] If you are relatively close to the Earth's surface, the value for g is approximately 9.8 meters/second squared (m/s^2). Gravitational acceleration can change, depending on where you are relative to the Earth. The farther away from Earth you are, the lower the value for g becomes. But even if on a jet airliner many tens of thousands of feet above Earth the value of g does not deviate too much from 9.8 m/s^2. At this point, you may have wondered how it is that astronauts can experience "zero-g" in space. Remember that gravity is also in the areas of space that astronauts are, so the zero-g that astronauts and cosmonauts feel is related to an interesting phenomenon we already

---

[10] Do not get confused between the "little g" (g) here and the "big G" (G) discussed in chapter 3 as they represent two different things. Little g is the local value of acceleration if left in free fall due to gravity and is therefore called the gravitational acceleration, whereas big G is a constant that never changes and is just a number that you multiply to Newton's gravitation equation.

discussed in chapter 2, called free fall. See Appendix B for more information on this zero-g.

Weight is a force just like gravity and points in the same direction as gravity does. On Earth, weight points downwards towards the center of the Earth.

In everyday speech, people use the terms "mass" and "weight" interchangeably, but they are obviously two different concepts. Related, but different. Mass never changes, but weight can. Mass is the amount of matter within an object and never changes, but weight is a force and is caused by gravity pulling on the mass, and can change.

We've discussed gravity and its mechanics, how it is related between bodies and their role in weight and free fall. There is still much to consider concerning motion, and that is where we'll return to in the next chapter.

CHAPTER 5

MOMENTUM

"A lot of prizes have been awarded for showing the universe is not as simple as we might have thought."

- Stephen Hawking

A Side Note on Terminology

Many physics terms have been co-opted and used in our ordinary colloquial speech. Their meaning can be identical, or close, to the scientifically used versions, such as the word "inertia." This is a physics term through and through, but its meaning is also used in our everyday speech in much the same way as found in science. Inertia, in both physics and colloquial usage, means a resistance to move or to change movement.

Sometimes, scientific terminology may be so far off when used in everyday parlance as to make the term appear like a completely different one from that used in science. A great example is the oft-used, but greatly misapplied phrase "quantum leap," such as when saying, "when the internet was first created it marked yet another quantum leap forward in our ways of communicating with one another." (Incidentally, this is one of those phrases that really irks me when spoken outside of a scientific context. Unless, of course, it is used correctly – you may have to wait until volume 3 when we talk about quantum mechanics to know the real meaning).

The word momentum can be categorized as an example of the former, as it is usually used to refer to an idea or object that is moving with a lot of mass and/or speed behind it. This can be in both an actual or in a metaphorical sense. Actual: "the cruise ship had so much momentum behind it, that it was difficult for the captain to get it to slow down enough to avoid hitting the dock" vs. metaphorical: "the novelist's works had so much mass appeal and momentum, that it contributed to their adaptations into plays and movies".

What is Momentum?

Obviously, in the world of physics, we speak of real objects and their motions. So, what exactly is momentum? **Momentum** is a concept of movement and has two components associated with it that, when multiplied together, gives you momentum. They are mass and velocity, each of which was discussed in Chapter 1.

> Traditionally, momentum is labeled as "p" when writing equations[11]:
>
> $$p = m\,v$$

Now, since velocity has a direction associated with it, momentum will as well after calculating it. This means that both velocity and momentum are vectors. When we say something has a certain momentum, we need to intuitively think that the object has an "amount of massive speed going in a particular direction". The more mass, the more momentum. The more velocity, the more momentum (and consequently, the longer the vector).

Unlike mass, which has a unique unit of kilograms associated with it, and time has its novel units of seconds, there is no uniquely named unit for momentum. Since it is defined as mass times velocity, momentum's unit is the combination of both: kg x m/s. For example, we would say that an object's momentum is downward at 10 kg-m/s.

---

[11] You may ask, why do we use the letter "p" for momentum? It originally comes from the German word "der impuls" and the French word "l'impulsion", both of which mean "pulse". The letter "I" in the beginning of these words could not be used, as it was already used to represent the different physics concepts of "impulse" and "moment of inertia". Obviously, the next letter, "m", could not be used either, since it was already in use for mass. So, the next letter in the German and French words was "p" and that is where it stayed.

Impulse

There is a physical concept in physics called **impulse** that becomes important when we speak about changes in time. Impulse is defined as the change in momentum. Remember, there are two variables associated with momentum: mass and velocity. So, if either mass or velocity or both quantities change over time, you are creating impulse.

The traditional example used in physics to discuss impulse is a rocket. A rocket starts on the ground. If there is enough thrust from the engines, the rocket begins to lift off. Gradually the rocket will accelerate or increase its velocity upward. Also, the rocket starts on the ground with a massive amount of fuel. However, to get that massive spacecraft off the ground, a lot of fuel will need to be burned. This means the gradual loss of fuel over time. So, what we are seeing is a prime example of an object that is both changing its velocity (going upward at an increasing speed) *and* changing its mass (burning its store of liquid or solid fuel). The rocket has a lot of impulse.

Another good example of impulse is that of a moving billiard ball colliding with a resting ball and imparting velocity to it. Initially, the resting ball has no momentum whatsoever, since its velocity equals zero. When the moving ball hits the resting one, that one now begins moving. So, the resting ball went from resting to moving, which is a change in velocity and therefore, a change in momentum. We would say that the moving ball gave the resting billiard ball an impulse.

I'm a Star Trek fan, and those of us who listen carefully will occasionally hear the captain tell the navigator after coming out of a warp to switch from warp engines to the "impulse drive." Presumably, these are the Enterprise's secondary and slower engines for regular cruising speeds. For now, full speed ahead...

Conservation of Linear Momentum

There are rules of nature that hold everywhere within the universe and are never broken, regardless of when they have occurred, or where they will occur. These rules are deemed "laws" because of this unbroken and universal character. One of these is called the **conservation of linear momentum**. It states that assuming no forces are acting on a system, the linear momentum before an event will remain the same after the event as well. The word linear means exactly that - straight lines or along a line. There are other forms of momentum such as rotational (which also have their conservation laws), but we will hold off explaining those until later. Let's explore how this conservation law works.

Using the simple example of billiard balls again, start off with billiard ball A that contains a particular linear (straight line) momentum of $p_{A(initial)}$ heading toward another billiard ball B that is currently at rest. Ball B has a linear momentum of zero ($p_{B(initial)} = 0$) since it is not moving. We can see that an event will soon take place where ball A will collide with B. So before this occurs what we need is to simply add up all the linear momenta of this whole system. This turns out to just be $p_{A(initial)}$ (since $p_{B(initial)}$ is zero, we can disregard it). This is called the "initial momentum of the system."

Next, ball A collides with ball B, and the result can be one of two scenarios: 1) both balls move or, 2) one ball moves. (You will never get a case 3 where both balls end up at rest, that would violate this conservation law). In both cases, the resultant total linear momentum (also called the "final momentum of the system") remains the same as before the collision – it needs to equal $p_{A(initial)}$. So, in case 1, when both balls are moving, they will each be moving slower than the velocity that ball A had before the collision. However, if you were to add both of their linear momenta together, they will equal the initial p_A. In scenario 2 when only one ball moves (this will have to be ball B as ball A has transferred all its momentum to the ball it collided with) then again, the final ($p_{B(final)}$) will be equal to the initial p_A.

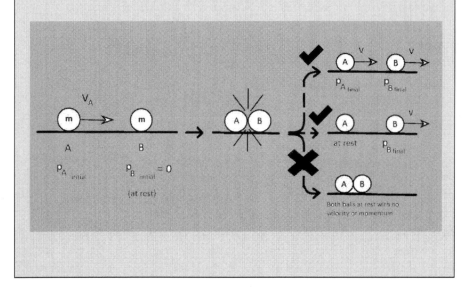

Yes, this can all be very confusing! Trying to juggle these different subscripts and very similar variables associated with different momenta can easily and quickly become convoluted, but the concept is simple to understand. The important takeaway point is that the total of all the initial momenta will equal the sum of all the final momenta and that is the conservation of linear momentum.

CHAPTER 6
ROTATIONAL MOTION

"You spin me right round, baby

Right round like a record, baby..."

- Dead or Alive

The Anatomy of a Circle

Let's discuss a geometrical figure that features prominently in physics and whose structure is fundamental to analyzing many physical phenomena: the circle. It's a beautiful thing! Normal rotations around a fixed axis (one that does not move), such as a wheel and axle, a merry-go-round, a Ferris wheel, a blender, etc, involve circles. So, we need to get familiar with these figures and their parts. A straight line from the very center of a circle out to the circle is called the **radius**. The **diameter** of a circle is twice the length of the radius and extends straight through the middle of the circle cutting it into two equal halves. The **circumference** of a circle is simply the length of the outside perimeter of the circle. A portion of the circumference is called an **arc**. If you make a wedge-shaped area, swept out with a radius out to an arc, this wedge is called a **sector** of the circle.

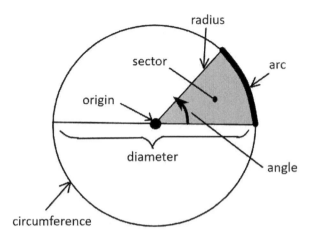

The anatomy of a circle

We've all seen the Greek letter pi (π), but what does it represent? It is defined as the circumference of any circle divided by that circle's diameter (π = circumference/diameter). Pi is a non-ending number that is roughly equal to 3.14. So, take a circle's circumference and spread it out into a straight line. Next, take that circle's diameter and measure how many of those diameters can fit along the spread-out circumference and you will get approximately 3.1415.... of them measuring the circumference. That is the definition of π.

If you move a radius along a circle like a clock hand, it will mark out angles, which are traditionally labeled with the Greek letter theta "θ." This measure is in either of two units: radians or degrees. A **radian** is the measure of the angle that is made when its arc along a circle is as long as the radius (if the arc were to be straightened out). There are 360 degrees (360°) or 2 pi radians (2π) in a circle. [12]

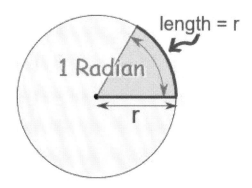

You can use either radians or degrees when figuring out angles, but you must be consistent with your other measures and calculations; you cannot mix the two together. All your calculations must be in radians, or they must all be in degrees and neither the twain shall meet. It would be like mixing American dollars and British pounds together.

Rotational Motion

So far, we've discussed linear motion, where objects move along simple lines. However, nature can be far more complicated than that, and we need to look at those cases where motion is rotational as well. **Rotation** occurs when an object or

---

[12] The two concepts of π and 2π can get a little confusing. As mentioned, π is defined as circumference of a circle divided by the diameter (π = C/d). However, when we say 2π we really mean circumference divided by the *radius* of the circle (2π = C/r).

group of objects turns or pivots around an axis. The axis of rotation can be within the object itself or outside of it. When the axis is outside of the object, its motion is called an **orbit**.

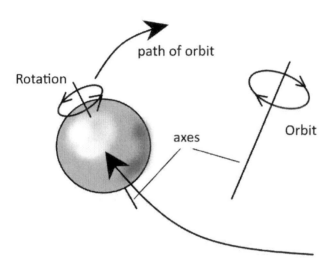

Difference between rotation and orbit. In rotation, the axis is within the object. In an orbit, the axis is outside of the object.

Just as there are linear concepts of velocity, acceleration, momentum, and force, there are rotational analogies to these in physics as well. Even the concept of mass in linear motion has a substantially similar notion in rotational motion.

An important concept to mention is that just as mass has a certain inertia associated with it, rotation does as well. Newton's first law states that objects in motion stay in motion, and objects at rest remain at rest unless an outside force is acted on them. This applies to both linear and angular movements. So, an object that is rotating will continue to do so, unless some kind of force acts on it to do otherwise.

Let's explore these ideas.

First, we'll need to imagine a solid disk, like a pulley or a wheel, that is rotating around its center at a constant speed. Pick a point somewhere on that disk and call it point P. The line connecting P to the center we will label as r, for radius.

When an object rotates about its center the point P that is r distance from the center on the object maps out a curved arc length, Δs. The angle that the rotation makes is called the angular displacement, $\Delta \theta$. All three variables are related as:

$$\Delta s = r \cdot \Delta \theta$$

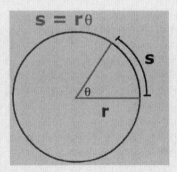

Velocity, as we know, is a measure of constant directional motion in a straight line. Its rotational analog is called **angular velocity** and is represented mathematically by the Greek letter omega (ω, the Greek equivalent to "v" in English). Now, that point P on the disk has two kinds of speeds associated with it – a translational velocity and an angular velocity. If there happened to be a little particle attached to P that suddenly detached from the disk, which direction would it go? It would go in the direction of the *translational* velocity. This is a standard linear velocity that is directed perpendicular to the radius.

Linear velocity (v), the radial distance from the axis (r), and the angular velocity (ω) are all related as:

$$\omega = \frac{v}{r}$$

Similarly, there is the analogous relationship between linear acceleration (a), radial distance (r), and **angular acceleration** (denoted by the Greek letter alpha, α, the equivalent to English's letter a), which is the change in the rate of angular velocity:

$$\alpha = \frac{a}{r}$$

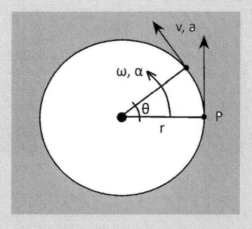

Visual differences between linear velocity and acceleration (v, a) on a circle and the rotational analogs of angular velocity and acceleration (ω, α).

In physics and astronomy, when we try to analyze a situation, we like to model reality in its simplest form. That means removing all extraneous details from the model that have no bearing or influence on what is being analyzed. When it comes to analyzing the motion of objects, one thing we can do is to portray the object as a point object instead of an extended object, if it is only the motion that we are concerned about. Every extended object has what is called a center of mass. The **center of mass** represents the point of an object where all its motional behavior appears to occur as if all its mass were concentrated at that one point. A spherical mass, like a bowling ball, with a constant mass distribution throughout it, or a spherical mass where most of its mass is concentrated closer to its center, all have centers of mass at their very centers. Other objects that are symmetrical around a line usually have their centers of mass on that line. Those objects that are asymmetrically or oddly shaped have their center of mass located somewhere else. In all these cases, the object can be thought of as being shrunken down to an infinitely small point, and only that point's location is what is important for motion.

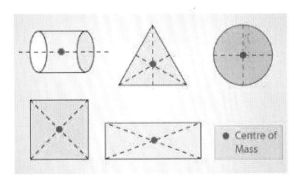

The center of masses of various geometrical figures

To give an example, if we were to throw a hammer at an oblique angle, the whole hammer would take on a usual parabolic arc through the air. However, it may spin around some point as it does so. It is that point that is called the center of mass, and it is also that point that moves in the parabolic arc.

The rotation of a thrown hammer about its center of mass. Notice that only the hammer's center of mass moves in the parabolic arc.

Center of mass is essential in the study of rotational motion because this is the point about which an object will rotate when forced to do so. When we want to find out how an object will move through the air or space, the center of mass is the point you use to make your calculations.

Torque

Imagine pushing on a ball that happens to be on ice (this is to preclude the influences of friction on the ball). If the force that you are directing onto the ball happens to be directed through that ball's center of mass, then the ball will move in a linear (straight line) fashion. If, however, the force that you direct onto the ball happens to be away from its center of mass, the ball will rotate.

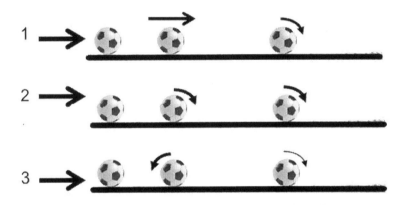

How an object moves on a surface when a force is applied depends on where the force is directed.

We're all familiar with using screwdrivers, wrenches, doorknobs, door keys, and pushing kids on a merry-go-round. These actions are examples of using a force called torque. **Torque** is the force applied to produce some sort of rotation around a pivot point. Don't get too confused by *pivot point* and *center of mass*. A pivot point is one that is usually fixed, such as the axle on a merry-go-round, that is attached to the ground or a doorknob attached to a door. Whereas, a center of mass is a kind of pivot point that is *unfixed* and is itself free to move about, such as the center of the Earth moving around the Sun or the hammer's center of mass when it traveled in a parabolic arc through the air.

We use the Greek letter tau (τ, the Greek equivalent to "t" in English) to denote torque in equations. In physics, torque is defined as the distance from the pivot point (d) the force is applied to times the perpendicular force (F) applied on the object:

$$\tau = F \cdot d$$

As an example of torque, let's use a wrench. We want to use the wrench to grip onto a bolt to unscrew it. The pivot point is at the center of the bolt itself. The distance d is that from the pivot point to your hand.

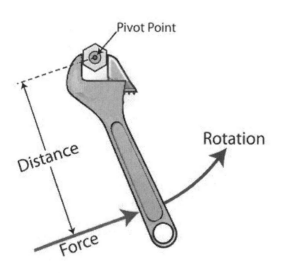

When you begin to turn the wrench with your hand, you are applying a force F. So, torque increases if the distance is increased and/or if the force used is increased. When trying to loosen a tough bolt a longer wrench applies more torque and that bolt will have to budge. As the ancient Greek philosopher and scientist Archimedes said:

> "*Give me a lever long enough and a fulcrum on which to place it, and I shall move the world.*"

All the previous examples mentioned (keys, merry-go-rounds, etc.) operate the same way. There is a force applied on the outside of the object a certain distance from the pivot point, and rotation occurs.

Now, torque is the rotational equivalent that a force is to a linear system. As before, when a force is applied to a mass, it produces an acceleration in it. This means the force makes the mass change its velocity, by either slowing it down, speeding it up, changing its direction or any combination of these. The rotational analog is that a torque produces an angular acceleration not to a mass, but to the **moment of inertia**.

What is a moment of inertia? It is the same concept as mass but is a quantity related to a mass in rotation. It is an object's *rotational inertia*. As discussed previously, an object resists rotational motion and will only do so when forced.

Earlier, when we spoke of linear motion, we were not interested in the shape of the mass, so we instead changed it into a point mass. To simplify our analyses of it, we introduced the center of mass to help shrink the mass all the way down to a single point. We cannot do this when looking at rotations, because when an object rotates, the object's *shape* now becomes important in the analysis. So, when looking at the angular momentum, angular velocity, and angular acceleration, we need to know the shape of the mass. Is the object a solid ball, a spherical shell, a rod, a barbell, a baseball bat, or some oddly shaped object? Since the object is no longer a one-dimensional point but is now a two- or three-dimensional object, when figuring out its angular equivalent of mass, we need to imagine its mass as tiny pieces or subunits all stuck together. Maybe the object is a disk, and its mass is equally distributed throughout the disk. Each subunit contributes to the total mass of the object.

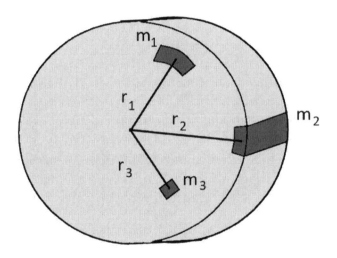

Selected random subunits within a solid disk. Each piece of mass has a unique radius from the center of rotation. The mass and the radius of each subunit give that unit's moment of inertia. The sum of these moments of inertia gives the entire disk's moment of inertia.

However, when figuring out the object's moment of inertia (I), you must take the subunit's mass times the distance from the center of mass squared ($I_{subunit} = mr^2$). You must then sum up all of these small subunit's moments of inertia to get the entire object's moment of inertia. There are multiple techniques in higher mathematics, such as calculus, that are used to sum up all the subunits, and every different shape may have a different overall moment of inertia. For example, the total moment of inertia of a sphere that is rotating along an axis that goes through its center is:

$$I = \frac{2}{5} MR^2$$

Where M is the sphere's total mass, and R is its outer radius.

A total moment of inertia of a cylindrical rod that is rotating around an axis through the center of its length is given by this equation:

$$I = \frac{1}{2} MR^2$$

As you can see, the shape of the object is crucial in determining its moment of inertia. That's not all, though. Where the object's rotation axis is located and in what orientation it is in, is also vital in calculating the object's total moment of inertia. Take the solid rod once again. This time make the rod rotate not around its long axis, but instead at 90 degrees to it through its center. Picture the rod rotating like the blades of a helicopter. In this case, its total moment of inertia is:

$$I = \frac{1}{4} MR^2 + \frac{1}{12} ML^2$$

Where L is the length of the rod.

It is evident that the shape and mass distribution of the object that is spinning will make all the difference in its moment of inertia. A solid disk with a constant mass distribution will be different than a ring with the same amount of mass, but the ring's mass will be distributed instead along the rim itself. In the case of the disk, each individual subunit contributes a little to the total moment of inertia; the subunits closer to the rotation axis (center of mass) will contribute very little because each of their mr^2 will be small (they are closer to the axis, and therefore the r^2 is small).

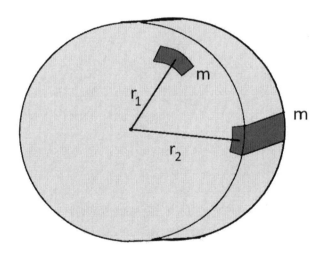

In this solid disk, we have highlighted just two subunits, each of the same mass. However, one is closer to the axis of rotation (r_1), whereas the other is farther away (r_2). The mass subunit that is farther away will contribute more moment of inertia to the disk's total since its radius is longer.

Those subunits closer to the rim of the disk contribute more to the moment of inertia due to their greater distant r from the axis. Since some of the subunits contribute a small amount and others provide more, these moments of inertia add up to an average value for the disk. The ring, however, is a different story. In the ring's case, all its mass is along the circumference and none within. So, every subunit of mass has a significant moment of inertia (each one's r is large and, therefore, its mr^2 is also large). Since the moment is generally larger in a ring of the same mass as in other comparable shapes, it is the shape of choice for objects that need to retain their motion for more extended periods of time, like flywheels in cars and in spinning gyroscopes.

Angular Momentum

We'll first need to know what angular momentum is. **Angular momentum** represents the rotational analog of linear momentum and can be thought of as spinning mass.

To get an equation for angular momentum, we must go to its linear equivalent, linear momentum, which is $p = mv$, and then replace the mass m and the velocity v with their rotational analogs. As discussed earlier, the replacement for mass is the moment of inertia, I. The velocity replacement is angular velocity ω. In physics, angular momentum is denoted with a capital L. So,

$$L = I \cdot \omega = (mr^2) \cdot (\omega) = m\omega r^2$$

Let us sum up all of the equivalent measures between straight line, or linear, motion with those of rotational, or angular, motions:

Linear Motion	Angular Motion
Displacement, x	Angular displacement, $\Delta\theta$
Velocity, v	Angular velocity, ω
Acceleration, a	Angular acceleration, α
Mass, m	Moment of Inertia, I
Momentum, p	Angular Momentum, L
Force, F	Torque, τ

Conservation of Angular Momentum

Now we are in a position to discuss the next conservation law: the conservation of angular momentum. Angular momentum can change in several ways. One is by changing the mass of the object rotating. Imagine the merry-go-round again. Children are on it, and they are spinning around the axis. If no one is pushing the ride around and no kids are kicking it forward, then the whole system has a certain angular momentum associated with it that will not change. The mass of the merry-go-round is the total of all the kids on it plus the mass of the ride itself. If one of the children were to jump off the ride, then the merry-go-round will obviously have less mass than it did before.

Doesn't this mean that the angular momentum will change? The answer is no, it will not, due to the **conservation of angular momentum**. Just as the conservation of linear momentum means *linear* momentum remains the same before and after an event (assuming no outside forces interfere), the same happens in this scenario. In this case, the event is the child jumping off the ride. Before and after the jump the *angular* momentum remains the same. Mass m obviously reduces after the jump, so something else in the angular momentum equation ($L = I\omega$) must change to compensate for this, and thereby bring L back up to its initial value. The variable that does change is ω. Omega must increase. Well, omega is the angular velocity. This means that when the child jumps off the ride, the ride spins slightly faster afterward. It has a new angular velocity to compensate. So, $L_{initial}$ must equal L_{final}. And *that* is the conservation angular momentum.

Playing with the conservation of angular momentum. The playground's merry-go-round starts off with a certain amount of angular momentum when the

children are on it, and it is spinning. That amount is
retained even when a child jumps off of it, but to
compensate, the merry-go-round must spin a little
faster.

Here's another classic example of the conservation of angular momentum. In this example we are not removing any mass from the rotating system, like earlier, but we are going to change the overall moment of inertia, instead. An ice skater starts off spinning on the ice with her ice skates, and she has her arms splayed outward. There is a certain angular momentum ω and, as a result, a specific angular momentum L associated with her spin. This is her initial state. Now, she brings her arms inwards slowly and finally hugs her body. This is her final state. What happens to the angular momentum? Well, her mass does not change, but her mass *distribution* changes; the mass subunits within her arms change their distances r from the axis of spin. So, some of the mass' r^2 have reduced and to compensate ω must increase again. L remains the same in the final state – conservation of angular momentum.

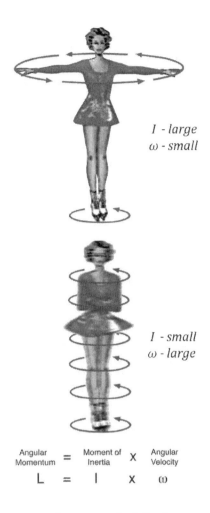

I - large
ω - small

I - small
ω - large

Angular Momentum = Moment of Inertia X Angular Velocity

$$L = I \times \omega$$

A spinning ice skater wonderfully demonstrates the conservation of angular momentum. The first image shows the skater with arms outstretched with a high moment of inertia, but spinning moderately fast. When she brings her arms in, she spins faster, due to a lower moment of inertia. This is to retain the same angular momentum throughout the spinning.

Previously, we were careful to note that there were no outside forces acting on the system. In the case of linear

103

momentum that would mean outside linear forces, like gravity, friction, air drag, and electric or magnetic forces. In the case of angular momentum, the outside forces we mean are torques, such as pushing or pulling on a wrench handle, twisting a doorknob, etc. The reason we must be careful in removing these forceful and interactive influences is that they introduce increases or decreases in angular momentum that *will* violate the conservation of momentum. *These conservation laws only work within isolated systems.* Let's see some examples of this violation at work.

Let us go back to the park with the children on the merry-go-round. The children are having fun standing or sitting on the ride in their same respective positions. Along comes a concerned parent and believes their child is too fragile to withstand the speed the ride is going at and wishes to slow it down. The parent slows the ride down by sliding their hand on the handlebars. What is occurring here is the introduction of an outside torque on the system of ride plus children. Can you guess what that external force is? If you said friction, you would be correct, and it is a torque on the merry-go-round that reduces its rotational speed. So, the initial L is no longer equal to the final L, and angular momentum is no longer being conserved. In the initial system, no friction existed; in the final system, friction was. The system was *not* isolated.

Another example would be the rotation of the Earth. It has a certain constant angular momentum associated with it. Along comes a large meteorite from outer space that impacts the surface at an oblique angle and leads to a dinosaurian dystopia. The new momentum of the Earth will change from its initial state, albeit in a very slight way. (For example, if the trajectory of the impact happened to be *against* the rotation of the Earth, then the Earth would have slowed down slightly afterward. If the converse occurred, and the impact trajectory was in the *same* direction as Earth's rotation, then the Earth's speed would have increased).

The impacting meteorite was an outside source of torque, and the system was not isolated.[13]

In this chapter, we discussed different aspects of rotational systems. We started with the anatomy of a circle and its component parts. This gave us a toolkit that allowed us to properly approach the analysis of rotational systems. This was applied to generic rotational objects, and then we asked what caused this particular type of motion. We found that torque is to blame. Next, we discussed angular momentum and how rotational systems have their analogs in the linear motional world. Lastly, we saw how angular momentum, just as linear momentum, conserves its properties, as long as they are found within isolated systems without torques applied from without.

When we think of physics, most of us think about motions, forces, and energy. Well, we've covered the former two concepts, our next point of interest in our physics express itinerary will be in one of physics' Holy of Holies; the land of work and energy.

---

[13] This is, however, a simplified model. In reality, a more complicated situation happens that actually retains the conservation of momentum *if* one were to initially include the meteorite within the model prior to impact. As long as the moving meteorite *plus* earth are included in the pre-impact analysis and the post-impact analysis the system remains isolated. Essentially, what needs to occur in that case is to resolve both object's momenta into their individual components. Then, by adding them up, pre and post, they will remain the same.

CHAPTER 7
WORK AND ENERGY

"The fundamental concept in social science is Power, in the same sense in which Energy is the fundamental concept in physics."

- Bertrand Russell

Work

Our everyday notions of work are not necessarily those shared by physicists. When we speak about work, we mean our jobs. We mean back-breaking and exhausting activities that may make us sweat. Any activity that takes any physical or mental effort is considered work in the colloquial sense. Physicists, on the other hand, have an exact definition for work and, of course, feature that esoteric language called mathematics. In physics, **work** is defined as the force applied to an object times the distance that the object moves *in the same direction* as the force.

Mathematically, this is expressed as:

$$W = F \cdot x$$

Where W is work, F is the force, and x is the distance that the force was applied.

So, if we push against a wall and push and push and eventually start sweating from doing so – well, sorry Bub, but no work! Yes, you were applying a force, but that wall didn't move. Carry a heavy weight and begin walking with it. Sorry Charlie, no work. Yes, you are carrying the weight and applying force upwards against the gravity, and yes, you moved the weight when carrying it. However, the movement of the weight was *perpendicular* (at right angles, or 90°) to the force's direction (upward). This means that no component of the force you used to keep the weight up was contributed to moving it forward. Push something forward (or pull something toward you) and make it move in *that* direction and you *are* doing work. That's the scientific definition for work, sweat or no sweat.

Walk up a flight of stairs and your legs push down onto the stairs, the stairs push back against your legs, and you move up. The stairs applied work on you. That is correct, the *stairs* did the work on *you*! (Which direction did you move? Upward. What force pushed you in that direction? The upward reactive force of the stairs against your downward forceful foot.) Lift some weights off the ground, and the upward pulling force applied by your hands moves the masses in that direction. Your hand is doing work on the weights. A car's wheel frictionally grips the ground and applies a backward force against it. Due to the rubber's friction with the ground, the ground supplies a reactive, forward, frictional force against the tires, which pushes the car forward. The ground applied work on the vehicle. A plane begins to rise to a higher elevation. Its wings are shaped so that air underneath them applies an upward force called lift against the wings. The air does work on the plane.

All the previous examples are of work using linear forces, but what about work made through rotational motion? Just as we did earlier with angular momentum, we replaced the linear variables with angular ones, we do the same here. Since $W = F \cdot x$ in linear motion, we replace force (F) with torque (τ) and displacement (x) with angular displacement ($\Delta\theta$):

$$W = \tau \cdot \Delta\theta$$

This is the work done to an object when a torque is applied to it through a certain amount of angular displacement.

Conservative and Non-Conservative Forces

We needed to have an idea of what work really was and how it operated before we could move on to an additional concept related to forces. There are a variety of forces and torques in the world, all of them pushing or pulling (or twisting) in one way or another and causing masses to accelerate. When a force happens in precisely the same direction (or in the opposite direction) of movement that a mass moves, we say that work is being done. But what about in those directions that are diagonal or perpendicular to the direction of motion? Is work being done then? If the force vector is pointing even in a slight way towards the direction of movement, then only that small portion in the direction of motion is contributing to the work being done. It is only when the force vector is perpendicular (or right angles) to travel that no work is being done.

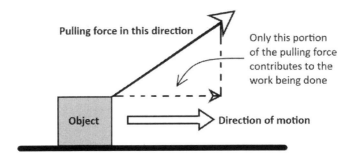

A force vector can be divided up into a couple of mutually perpendicular vector components (the dashed lines). Only the component that is parallel to the direction of motion ever contributes to the work being done on the object.

This brings us to the categories of forces in nature. There are two primary flavors of forces in the universe, which represent the conservative and nonconservative forces.

Imagine an object starting at point A. There is a second point, point B, farther away that the object travels to. It may go along any path, as long as its starting point is A and its endpoint is B. Now, imagine a force that the object feels along the way. There will be work done along the path. The question is, does the amount of work done by the force depend on which path is taken? If the answer is yes, then you are dealing with a **nonconservative force**. In other words, the work done is "path-dependent." On the other hand, if the work that is done remains the same no matter what path you take from A to B, then it is a **conservative force**. A conservative force does work in a "path-independent" way.

Allow me to illustrate the difference a little more clearly, as this idea can be a bit confusing. You are at the bottom of a small mountain, and you plan to make a day hike to the summit.

There are two trails to choose from. Trail 1 will take you from the bottom (point A) to the summit (point B) in the most direct, straight, and shortest way, but the path is very steep. Trail 2, on the other hand, will still take you from the bottom to the summit but will do so in a much gentler, flatter, but longer path. Trail 2 will wind its way in a curvy ascent to the summit. You have the force of gravity as your friend to continuously remind you of the work you are doing against it. Question: which of these two paths will you do the most work against gravity? Answer: both paths will need the *same* amount of work done in going from points A to B. *Therefore, gravity is a <u>conservative force</u>.*

The reason this is so, is because gravity is directed in only one direction – downward. There will only be work done against it when there is movement in the upward direction. Movements to the left or to the right make no difference, and you can make as many of these movements as you want without doing any work. There will be no contribution to work against gravity if you move to the sides. When you walk through trail 1, you are going in an upward direction the entire way, and therefore doing work against gravity the entire way, but it is shorter. When moving

110

through trail 2, work is being done against gravity only when there is an upward movement. However, there is precisely the same height between points A and B regardless of which path is taken (both routes have the same beginning and ending points) and, therefore both paths do the same work. Other examples of conservative forces are those from electric and magnetic forces and from elastic forces, such as from springs.

Let's look at an example that demonstrates a nonconservative force. Your young son wants you to pull him on a mat across the carpeted floor from the living room to the kitchen. There is expected to be quite a bit of friction between the mat and the carpet (and maybe a few sparks and some hair standing on end, but we'll disregard those details). You decide to minimize the amount of work you need to put into the effort. Does it matter if you go straight from the living room to the kitchen or do you take a detour through other rooms before getting to the kitchen? It's pretty obvious to us fathers that more sweat will be involved in the second method of travel; the direct path is the one of least resistance; the longer path leads through and against more carpet travel. The amount of frictional work is path-dependent and so, *friction is a nonconservative force*. Other than friction, those forces caused from resistances through fluids, such as air and water drag, are yet more examples of nonconservative forces.

Kinetic Energy

The science of energy takes on a world of its own. It may not be an understatement to say that energy is the center of the physics universe. Vast fields and industries are devoted to the utilization of the different forms that energy can take. **Energy** is an abstract concept and is usually defined as the ability to do work on an object. This is the reason why we needed to discuss work before dealing with energy.

Some examples of the many forms of energy are heat, electricity, magnetism, mechanical, potential, and nuclear, among others.

We will discuss mechanical and potential energies in this chapter. **Mechanical energy** consists of the energy inherent in movement or location. Mechanical energy comes in two forms: kinetic and potential. **Kinetic energy** is that found in motion, and **potential energy** is the form taken from the position an object has relative to a reference energy or point. We will discuss potential energy in a later section in this chapter.

Imagine a baseball pitched through the air. At any moment in time, the baseball has a certain velocity associated with it and always has a constant mass as well. As discussed in an earlier section, the baseball has a momentum due to its mass times velocity (p = mv), but these two variables form the components to the baseball's kinetic energy, too. The equation for kinetic energy is:

$$KE = \frac{1}{2}mv^2$$

KE represents kinetic energy.

In the SI system of units (metric), all energy has the units of Joules (J), named after the 19th century English physicist James Prescott Joule, who found relationships between heat and mechanical energy. To get an intuitive feel for the equivalency of 1 Joule of energy, hold a medium-sized apple in your hand and raise your hand up about 1 meter. You've performed about 1 Joule of energy moving that apple against gravity.

Work-Energy Theorem

Okay, so we have an object that is moving with a constant speed, but what if we find that it changes its speed; what causes this? As we learned in an earlier chapter, which discussed Newton's laws of motion, we found that his first law stated that any object in motion continues at a constant speed, unless acted upon by a force. If the object starts to accelerate, it has to be caused by a force. Remember also from the previous section that when a force acts along a certain distance on an object, it is doing work on that object. Well, that is what is occurring here. The force vector and the motion vectors are in the same directions. The force is acting with the motion and causes the object to speed up as a result. Because there is a change in velocity, there is also a change in the kinetic energy. Over the distance that the object changes its motion, the force is doing work on it. We now have a relationship between work and kinetic energy:

$$\text{Work} = \text{change in kinetic energy} = \Delta KE$$

This equation is called the **Work-Energy Theorem**. We'll soon find that even potential energy can be associated with this equation. So, now we know that whenever an object changes its motion, work, in the scientific sense, is also being done.

Potential Energy

A skydiver gets suited up and seats herself in a passenger chair within a small plane and begins the ascent to the sky looking forward to her upcoming jump. As she rises, she is unknowingly placing currency in the bank of her potential energy fund. The plane finalizes its ascent, levels off, and her potential is filled to full capacity. The skydiver jumps and free falls toward the Earth with exhilaration. Little does she know that as she falls, she is using up her fund of potential energy.

While kinetic energy is the energy of movement, potential energy is stored energy. There are different forms of potential energies, and all relate to environments that provide some sort of force or objects with pent up energy. There are gravitational, magnetic, electric, nuclear, and spring potential energies. Fuels can be created that use stored potential energy. Other familiar forms of potential energies are solar, wind, water, natural gas, and fossil fuels. Then there are the objects created for destruction, such as dynamite, gunpowder, and nuclear bombs.

The study of potential energies is fascinating. The applications are endless and powerful. Civilizations have risen and fallen throughout history based upon their reliance and accessibility to potential energy sources. By lighting, sparking, dropping, compressing, or releasing these energy sources, they are transformed into forms of kinetic energy that can be utilized.

The earliest usage of potential energy came from burning wood. Plants absorb incoming sunlight and carbon dioxide in the air and funnel the energy and molecules into a biochemical process called photosynthesis whose end products are carbohydrates. These carbohydrates are stored within the plant for their own metabolism and for their structure such as in wood cellulose. Carbohydrates are large molecules that have potential energy stores that can easily be used by burning. Heat and light are the useful products of this burning, and our ancestors took advantage of it. Later in history came the advantageous burning of oils, coal, and natural gases.

Water is another source that was eventually used for its potential energy. Water's potential energy is stored in its height from a reference point, such as sea level. This is due to gravity's influence on it. When dropped, its change to kinetic energy is utilized to do work. This is most clearly apparent with a hydroelectric dam where the water level behind a dam is artificially kept at a much higher level than found on the other side. When water is controllably released from its higher level and flows through the dam, the weight and kinetic energy of the falling water spins turbines that generate electricity.

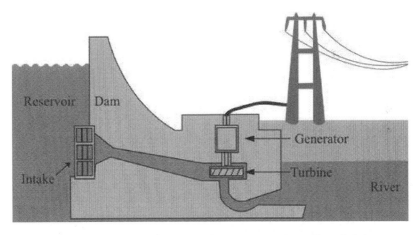

The height of the water in the reservoir (usually a lake) behind the dam gives the water potential energy. The water flows (with kinetic energy) through the dam,

spins the turbine. This generates electricity, and the
water continues on into the lower river.

Another utilization of water's potential energy was with milling equipment. Ancient millers would have to take grain and grind them to flour by hand or by work animal. This was done by rotating massive round stone disks upon each other with the grains between the two stones. Medieval millers devised a way to take advantage of water and would purposely build their milling equipment near streams and flowing rivers. This would allow falling water to mechanically turn the stones by directing the water to paddles that were attached to the stones. Water was used instead of human or animal muscle and endurance.

The water mill utilizes a higher level of water in a
river. This turns the water wheel, which in turn spins

machinery that grinds (mills) the grain. The water in
the wheel drops to a lower level in the river.

Anything of mass that feels a local gravitational field (such as Earth's gravity) has a weight associated with that mass. As discussed in chapter 4, weight is equivalent to mass times the gravitational acceleration (m · g) in the vicinity. I say vicinity because g can be different depending on where you are. Are you on Earth, on the Moon, in space near a planet, in deep space away from planets, etc.? The value for g depends on the local gravity of the area, but near the surface of the Earth, its value is approximately 9.8 m/s². Potential energy from gravity comes from moving an object away from one point in a gravitational field to another point that has a different strength in the field. When you do this from a location that is stronger to one that is weaker, such as lifting an object from the ground to a height, you increase the potential energy of the object. If you move the object to that weaker spot and release the object, it will release its potential energy into kinetic energy by moving toward the area of higher gravity. Move weighty water up a mountain and release it. It has stored gravitational potential energy at the top. The water will flow down the mountain in a rush of released kinetic energy.

Mathematically, we say:

$$PE_g = mgh$$

That is, the gravitational potential energy (PE_g) equals mass times gravitational acceleration (g) times height above a reference point (h), which can be any point you choose. This reference is traditionally the surface of the Earth. Since potential energy is relative to that chosen reference level, gravitational potential energy is a relative value; it varies depending on where the object is located.

Another resource that stores energy is wind. Wind turbines utilized air movement to rotate their blades to spin milling and other machinery, but nowadays they are used to generate electricity. Sailing ships and mariners obviously used air currents for ship movement. These air movements are obviously not potential, but kinetic, energy. So, how does air *potential* energy work?

Air must move to form wind for us to fully utilize its power; still, resting, air does nothing. Cold air is denser and so tends to fall, while warm air has the opposite properties – it is less dense and rises. The natural movement tendencies of these bodies of air set up currents and wind. The driving force for these movements is the Sun. It is what warms the air in the first place. The potential energy in the air comes when you have significant differences in temperatures between bodies of air. When these different masses of air are separated, you increase their potential energy. Or, alternatively, if the difference in temperature between the two masses becomes greater, then their potential also increases.

Another form of potential energy is in the form of fossil fuels. The Industrial Revolution ushered in the massive exploitation of these natural resources. Coal, natural gas, and petroleum are considered fossil fuels and are the primary global energy sources today. They are so-named due to their prehistoric origins. Most fossil fuels are derived from dead organic matter, such as the detritus accumulated from swamplands and forests, that have been buried, pressurized, and heated over millions and millions of years underground. Like dead and dried plant material that can easily be burned in a campfire, fossil fuels are extremely concentrated forms of this material that have also undergone chemical changes. These materials have their potential energies retained in the form of chemical bonds. When broken, these bonds release heat, light, and smaller molecules. Fossil fuels are used in everything from internal combustion engines within car motors, boilers to power trains, coal-burning power plants, gas ovens for home cooking, water heaters for warming tap water at home, to foundries and furnaces for producing and smelting ceramics, bricks, steel, and other metals.

The elasticity of springs can hold potential energies as well. Coiled metal has a stable configuration, but when squeezed together or pulled apart, there is an opposing force that increases with the stretching or the squeezing of the spring. The material of the spring tends to revert back to its more stable equilibrium point. This pent-up force is the potential energy. When the spring is released, that potential energy transfers to kinetic energy and the spring eventually finds its stable state again. Spring potential energies are used in wind-up clocks, pendulum clocks, vehicle wheel suspensions, floor, and wall absorbers, etc.

A spring's stable equilibrium point is considered its reference from which its potential energy can be calculated. When pushing or pulling a spring a certain distance x from that reference point, a force F_s is felt directed toward the reference point. F_s is also called a **restoring force**, since the spring wants to be restored to its equilibrium point. The farther away from that reference point the end of the spring is, the more force is directed toward that stable equilibrium point.

The force equation for a spring is also called **Hooke's Law,** named after the 17th-century British physicist Robert Hooke:

$$F_s = -kx$$

The constant "k" is called the **spring constant,** and the minus sign in front of the equation means that whatever direction x is from the equilibrium point, the restoring force will be in the opposite direction. The spring constant is a number that is found experimentally by testing the spring's elasticity. The larger the number that k is, the greater the spring's restoring force is. The material that the spring is made from, the number of windings the spring has, the thicker the spring's material is - all of these contribute to a specific spring's constant value.

Using the same two values in the spring's force equation (spring constant k and distance x from the equilibrium point), you can determine the spring's potential energy thus:

$$PE_s = \frac{1}{2}kx^2$$

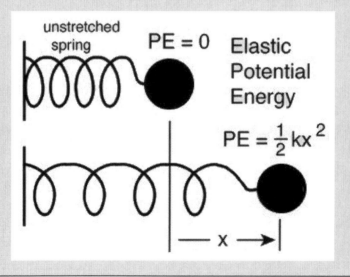

120

However, it must be said that a spring can be stretch only so far. In reality, if a spring gets pulled too far, its spring constant will change, and its equilibrium point will also change. The material the spring is made from gives way after a certain amount of stress on it until it breaks or deforms.

Conservation of Mechanical Energy

When an object has a store of potential energy, that energy can be converted into kinetic energy and vice versa. Think of an hourglass with two glass bulbs. Call the bulbs K and P, one of which is fully filled with sand. The sand in the hourglass represents some object's total mechanical energy. Each bulb represents a different form of mechanical energy, either kinetic or potential. Let us start with the filled bulb P and call it the potential energy bulb, while bulb K is the kinetic energy bulb. If we tip the hourglass over, the sand will start to drop into the opposite bulb. In our analogy the sand represents how much of the respective form of energy the object has. So, if one bulb, let's say bulb P, is entirely filled this means that the object that the hourglass is representing is entirely filled with potential energy and has no kinetic energy. In other words, the object may not at all be moving, but located in a place of maximum potential.

In some real-world examples, this may mean a pendulum that is at its widest swing and is momentarily still before reversing its swing. It may also mean a spring that is entirely squeezed or fully stretched and can go no farther. Maybe water is at the very top edge of a waterfall about to free fall. Another is a ball that has been thrown up and is now at its peak before beginning to fall. These are examples of one form of mechanical energy – that of potential energy – that happens to be at its maximum.

Bulb K, on the other hand, represents just the opposite: a container of kinetic energy. If this end was entirely filled with sand, we are at maximum kinetic energy and zero potential energy. Additional examples of this extreme are:

- A pendulum at the lowest, but fastest, point of its swing,
- A spring that has reached its stability (reference) point, but is passing through it at maximum speed during vibration,
- Water from a waterfall that, at the position just before hitting the ground, or its lowest point, has its maximum speed, and
- A thrown ball coming down and reaching its lowest point at maximum velocity.

All energies between these two extremes are mixes between potential and kinetic energy. The object under consideration will then have some kinetic and some potential energies associated with it.

The point of the illustration was to show that there is a certain fixed amount of total mechanical energy (all the sand in the hourglass) to be distributed. None of the sand is removed from or is added to the hourglass. So, total mechanical energy is constant and does not change. This is called the **law of the conservation of mechanical energy**. This law is usually stated as "energy can neither be created nor destroyed, but only converted into one form or another." Now, as with any conservation law in physics, the system we are looking at must be isolated. This means that no energy, nor forces, can enter nor leave from the isolated system that we are studying – the waterfall, the pendulum, the spring, the ball, etc.

Add up all the sand in both bulbs of your hourglass, and you'll get a fixed amount that never changes within it; it is conserved. So, the sum of potential energy and kinetic energy is a constant:

$$KE + PE = constant$$

Another relationship between potential and kinetic energies that can be expressed in a simple equation is the fact that whatever change in value (denoted as the Greek letter delta, Δ) happens in one, brings about the exact opposite change in value in the other:

$$\Delta KE = -\Delta PE$$

For example, if there is an increase of 200 joules in kinetic energy of an object, there will be a 200-joule *drop* in its potential energy.

Also remember from earlier that we have a very important connection between work and kinetic energy called the Work-Energy Theorem. But now we have a connection between kinetic and potential energies. So, here's an additional relationship to our work-energy theorem:

$$W = -\Delta PE$$

So, not only is work the same as the change in kinetic energy, but it is also the same as the negative of the change in potential energy.

Power

This brings us to our last concept in work and energy: power. **Power** represents the amount of work that is performed over a certain amount of time.

It is considered a rate and is expressed as:

$$P = \frac{W}{t}$$

W is the work performed (in units of Joules, J) and t is the amount of time it was performed.

The SI unit for power is the watt (W)[14], equivalent to a Joule/second, named after James Watt, the 18th-century Scottish chemist, engineer and inventor of steam engines. You may sometimes hear power expressed in the units of horsepower (hp). This is not a metric unit but historically was related to how much work a horse can pull in 1 second. 1 hp is equal to 745.7 watts. (Let's not get into the size, or the breed, or the sex, or the age, or the environment, that the horse happens to be in...just go with it!)

If you push an object with a force of 1 Newton a distance of 1 meter for 1 second, you have performed 1 watt of power. Lift a medium-sized apple a little over 3 feet and take 1 second to do it. You have used 1 watt of power.

In this chapter, we discussed some of the most important concepts in physics: energy and work. Not only are they restricted to physics, but they extend to almost every other branch of science. We talked about the differences between conservative

---

[14] Try not to get confused between the W representing "Watt" and the W representing "work". The former is the unit of power, whereas the latter if the symbol for work. That being said, yes, this can get confusing.

and nonconservative forces, kinetic and potential energies and the conservations of them. We found that work is done when a force is applied over a distance and that power is work that is done over some time.

We are almost complete in our discussions of motion. Presently, we've covered linear and angular motions. In the next chapter, we will be moving on to our final form of motion. That is periodic motions, also known as vibrations or harmonic motion. Onward and upward!

CHAPTER 8
VIBRATION

"I like the beauty of physics, nature is beautiful, it is my second soul to music."

- Fabiola Gianotti

Simple Harmonic Motion

The world presents us with many different varieties of motions. We've already encountered constant linear motions in chapters 1, 5, and 7. Non-constant accelerating motion, such as free fall, was discussed in chapters 2 and 3. Another kind of movement we've seen was found again in chapter 6, in which we learned about rotational or angular motion. There is yet another category of motion to probe, and it is a fascinating one. Those are the motions associated with repeatability or periodicity. Rotating motion is one case where this can occur. Ripples on the surface of water passing a single point give an example of repeatability. Springs repeat their motion when vibrating back and forth. A pendulum swing is another example of repeatable motion. These will be discussed in their respective turns.

When we speak about a repeatable motion, we really mean what is called a **simple harmonic motion**. This occurs when an object starts in one kind of configuration, begins a movement, returns back to its original configuration, and begins the entire process all over again. The simplest approximation we have to this kind of movement is from a model consisting of a

126

single point on the circumference of a circle that is rotating around the center at a constant speed. This is called **uniform circular motion**. If we want to time the speed of the point on the circle, there are many ways to do so. Some were already discussed in chapter 6 in the section on "Rotational Motion". They can consist of radians/second or degrees/second. Another handy measure could even be rotations per second. Regardless of how we measure the speed at which the point rotates around the circle, the analogy to other forms of simple harmonic motion is made this way. Take the point on the circle and project its current location horizontally as shown in the diagram below:

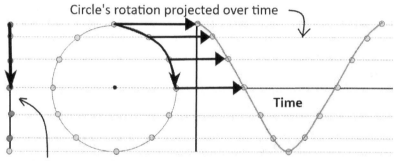

Circle's rotation projected over time

Time

Circle projected onto a wall sideways

Different kinds of simple harmonic motion (SHM) can be represented by a rotating circle's projection. The constant rotation of a circle is one form of SHM. Another form is an up and down motion on a line, like that shown on the left and which can represent a bobbing weight on a spring. This linear SHM is actually just a sideways projection of the circle's motion. Another form of SHM is that represented by a wave, like that seen to the right. This can be a vibrating string, a water wave, a sound wave, etc. All of these are the circle's projection to the right as time moves rightward.

The location of the point, when projected horizontally, will be the location of the point on the object that is performing the harmonic motion. This may be a block that is suspended by a spring, and that is undergoing a repeating up and down motion. The point may be on a fixed location coinciding with waves that are passing that location.

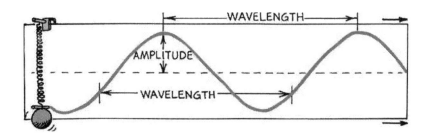

Like a simple circle, a wave has some terminology associated with it that we must learn before we can understand its motion. A single **wave** represents one entire cycle of motion. The **wavelength** (represented mathematically by the Greek symbol lambda, "λ") of a wave is the distance between two identical points in its cycle. For example, the distance between a peak and an adjacent peak is a wavelength. A wave contains the highest point in its cycle called a **peak** (or **crest),** and the lowest point in its cycle called a **trough**. The **amplitude** of a wave is defined as the height from the very midline between a peak and the adjacent trough up to a peak. The **frequency** (represented mathematically by "f") is a single number representing how many waves pass a point in a second and is measured in Hertz (Hz). The higher the frequency the more squashed together a collection of waves are and vice versa. So, if we say a particular sound has a frequency of 3000 Hz, we mean that every second there are 3000 waves passing by. However, the **period** of a wave is the amount of time it takes for one full wave to pass a given point. The period and frequency of a wave are inversely proportional to each other; when one increases, the other decreases, and vice versa. The idealized kind wave that bobs up and down as it travels in the right or left direction is called a **transverse** wave.

After analyzing this motion, you may ask: what causes an object to make a repeatable motion like this? As you can see, all of these points are either changing their directions or changing their velocities or both – all of which are accelerations. What causes accelerations? Forces do! There are certainly forces involved in every one of these motions.

In the spring and block examples, we learned about the spring's restoring force in which a force continuously builds up and points in the opposite direction of motion (toward the middle resting point of the spring, called the **equilibrium point**). This restoring force is what causes the block to slow down, stop, then reverse its motion and to accelerate toward the equilibrium point again. The restoring forces of springs come from the shape and rigidity of the material that makes up the spring.

In the case of water waves, the forces are a bit more complicated but involve the water molecule's cohesion with each other through their mutual electrical forces, and their interaction with passing pressure differences within the water causes little areas of water to rotate around a central point.

We will discuss the pendulum in more detail in the next section.

The Pendulum

A pendulum is another excellent example of simple harmonic motion and, of course, involves forces that explain its movement as well. A pendulum is simply a cord that hangs freely and has a mass attached to the end of it. This mass is made to swing from side to side in harmonic motion. If left alone, the pendulum eventually rests vertically at its equilibrium point.

When a pendulum is pulled away from its equilibrium point, a restoring force is acted on the pendulum's mass. The farther from this mid-point, the higher the restoring force upon the mass. How does this restoring force come about? It is formed from the mass' weight. The weight force is pointing straight down toward the Earth. When the mass is pulled away, that weight force has a small portion of it pointing along the arc path that the mass moves along toward the equilibrium point. It is this little portion that contributes to advancing the mass back to its resting point. As the mass moves along its pendulum swing, the part of the weight that points toward and along the arc will either get longer or shorter depending on how far away from or how close to the resting point the mass is located, respectively. In the diagram below, the pendulum's restoring force is labeled "F_t" (tangential force) and is a component of "mg" or weight.

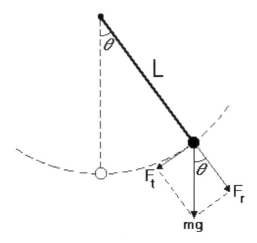

An interesting phenomenon concerning pendulum motion was discovered by Galileo Galilei during his experiments. He found that given a fixed length for the pendulum if you were to change the amplitude (that is, the distance the mass is moved away from its equilibrium position), it does not alter the pendulum's period. In reality, this is only an approximation and occurs with small amplitudes, not larger ones. Pull the pendulum away from its equilibrium point a little and let go. It will swing with a specific period. Say, one swing every 2 seconds. Pull the pendulum away even farther and let go. You will find that it will swing with the same period (again, every 2 seconds).

Period of Vibration

All objects that exhibit simple harmonic motion have an inherent rate of repeatability associated with them. This is called their **period**, T. The full back and forth swinging time of a pendulum and its subsequent return to its starting point is a pendulum's period. The time it takes for a stretched or squeezed, and released, spring with a mass on it to return back to that starting point is its period. Again, waves have precisely this same characteristic associated with them as well. They undulate with peaks and troughs that repeat over time. Their period is defined as the time it takes for one complete wave that includes one peak and one trough.

A handy list of reference period equations for some select objects of simple harmonic motion follows (this assumes small amplitudes):

Pendulum: $T = 2\pi\sqrt{\dfrac{L}{g}}$

Spring (vertical or horizontal): $T = 2\pi\sqrt{\dfrac{m}{k}}$

Wave: $T = \dfrac{\lambda}{v}$

Some variables in these equations need a little explaining. For the pendulum, L represents the length of the cord, while g is the local gravitational acceleration (9.8 m/s^2 in our familiar earthly environment). As can be seen, a pendulum's period does not depend on how heavy or large the mass is, or how far you start away from the equilibrium point (again, only for small amplitudes).

The spring's equation only depends on the mass of the object on the spring (m) and the spring's particular constant (k). This time, it does not depend on the gravitational acceleration.

Finally, a wave's period depends on the wave's wavelength (Greek letter lambda, λ) and its velocity, v. The wavelength of a wave is defined as the distance between two identical spots on a wave, such as peak to peak, or trough to trough.

While period measures the time that one full cycle of harmonic motion occurs, there is also the opposite measure of how many cycles pass in a certain amount of time. This characteristic is called a frequency (f) and is simply the mathematical inverse (flip) of the period:

$$f = \frac{1}{T} \ \text{Or} \ T = \frac{1}{f}$$

The unit for frequency is Hertz (Hz), named after the late 19th-century German physicist who proved the existence of the electromagnetic waves associated with light. So, 15 kilohertz (15 kHz) means 15,000 waves crossing a given point every second. Alternatively, this same wave will have a period (T) of one wave passing every 1/15,000 of a second.

Sound, Sonar, and Seismology

Can you hear those low-frequency rumblings that those fluffy clouds up in the sky above your head are making? No? Elephants can. Can you hear that extremely high-pitched chirp that those bats are creating? No? Their tastiest morsels of food, those lowly moths, can. Can you hear those strange songs in the ocean that those whales are making? Yes? How about their peculiarly detailed messages? Whales and dolphins can. Can you hear that vibration of the crystal inside your alarm clock while it's ticking away? What about those vibrations being produced from the termites within your walls? No, on both counts? Well, your dog can.

However, humans do have the ability to produce instruments that *can* hear all these sounds and more. This all begs the question: what is a sound? A **sound** is simply a vibrating wave that travels through a fluid or a solid medium. So, any state of matter can have a sound that travels through it.

Whatever mechanism that fabricates the sound is something that violently shakes the nearby particles near it. These particles vibrate, and their vibratory energy emanates outward all around the source in a concentric sphere that gets larger over time. When you look at a drop of water that falls onto a still pool of water, you'll see multiple waves spread out in concentric circles from that point. This is a two-dimensional representation of what happens with sound. Imagine adding one more dimension to this scenario, where there are no longer circles, but instead are round spherical shells spreading out in three dimensions from the source of a sound, like expanding bubbles.

In the case of water waves, it is the water medium that is rippling and undulating and is caused by the source droplet. With sound, the source is something that creates a very high-frequency shake. A tuning fork is a great example. Tap the tuning fork, and both arms of the fork will vibrate with a particular frequency. The arms are essentially shaking back and forth at a very high rate of speed. If we were to dramatically slow down the movement of the arms and were able to see the particles at an atomic scale, this is what we would see. With every forward acceleration of an arm, it collides and bunches up a large number of particles that the tuning fork happens to be immersed in, such as air. This piled up collection of particles is considered a denser part of the medium, and we call this portion of the cycle **air compression**. When the arm of the tuning fork reverses direction and starts moving back, then that same side of the fork is now moving farther away from the air particles, and the particles get spread apart. This portion of the medium becomes thinner and less dense, so it is called **air rarefication**. This process continuously repeats, and a collection of air compression and air rarefication waves spread outward from the tuning fork arms in a spherical wave. However, it is important to note that the only movement that the individual air molecules make is to simply jostle back and forth. They are not the things that are moving outward in an expanding bubble, only the wave of compression and rarefication does. The particles themselves stay fixed to their small local areas.

The waves can be thought of as those expanding bubbles of sound centered on the forks. Sound waves are of the **longitudinal wave** variety. This means that the periodicity of the waves happens in the direction of travel – the line of compressions and rarefactions is in the line of their travel.

As the waves spread outward, their intensities (energies) decrease. Think of the bubble's surface as being very thick at the beginning, but then gets thinned out as it gets larger. The thickness of the bubble represents the sound's **intensity**.

The tuning fork has a unique frequency associated with it, depending on the length of its forks, all other things being equal. The longer the arms, the lower the frequency, and vice versa. If the arms are relatively shorter, they will vibrate at a faster rate and therefore release a higher pitch. What this translates into is that the air wave's compressions and rarefactions are less spread out.

The speed of sound depends on the material it is passing through. If through the air, its speed will be about 346 m/s. In seawater, sound travels at 1531 m/s. The rate of sound in glass is 5640 m/s, but can be as high as 12,000 m/s through diamond! Generally, the speed of sound increases in media, the closer that the average distance the particles are to each other. So, sound will travel fastest through solids, less so in liquids, and slower yet in gases.

Sound's speed is not just affected by the medium it travels through, but also the temperature of the media. The higher the temperature, the greater the average speed of the composite particles, and the quicker the sound through it will be. Air at 0°C passes sound through at 330 m/s, whereas at 40°C, the sound will travel at 355 m/s.

We are all familiar with sounds in the air, and many are also familiar with those underwater as well. The use of sound underwater for measuring purposes is called **sonar**. It is an acronym that represents "sound navigation and ranging." It is a technology that was initially developed for detecting objects in the ocean for military purposes, especially for use by submarines. Pulses of high-frequency sound are emitted by a sonar source. These sound waves travel through the water and spread out. When a portion of the waves collide with an object, they get reflected back to the source, and a detector can "hear" these reflected waves. The sensor also measures the time from emission to the reception of the reflected waves, and this gives a distance to the object. We know the speed of sound underwater, so the longer it takes for the sound to make its way to the object being investigated, and to get reflected and received, the farther away the object is. This is where the term "ranging" comes from in the longer name for sonar. The "ping…ping…ping" sounds from sonar are the pulsed sound waves being released from the source, and the little bright blips that show up on radar screens are the objects that the sound waves have reflected off of.

A similar concept to sonar can be used through solids as well. For example, the solid earth can be "sounded" through the use of traveling pressure waves. The whole science of **seismology** is based on this principle. A violent shake at the surface of the Earth, or underground, causes waves to travel through the layers of our planet. These shakes can be artificially set up through the use of explosives, like dynamite, nuclear bomb tests, or from thumper machines that pound the ground like large hammers. (Remember David Lynch's 1984 sci-fi movie version of Frank Herbert's "Dune"?)

An earthquake is the prime natural example for producing strong underground waves. An **earthquake** occurs when two moving crustal plates get stuck against one another, and their pent-up pressures force the two to eventually break that sticking point. An enormous rumbling and jostling of the material of the ground spread from that point in concentric spherical waves. Again, the material of the Earth itself stays put in its area, but the wave of compression and rarefication is the entity that travels. The point on the ground directly above that rumbling source is called the **epicenter**. (The underground center of an earthquake is called its **hypocenter**). These so-called seismic waves can vibrate in two different ways, and seismic detectors can pick up both and distinguish them. One is called a P-wave (for primary wave) and is the faster of the two that travel. **P-waves** are those that are of the longitudinal wave variety, where the waves are composed of the periodic forward and backward compressions and rarefactions of the solid matter being traveled through. In other words, P-waves squeeze and spread the rocks in the *same direction* that they travel in. The other, slower, type of wave that travels through solids is called the S-wave (for shear wave). **S-waves** do not squeeze and pull rocks, but instead shake them up and down, left and right, or any combination of the two. The rocks shear, slide, or rub, against each other. More accurately, their vibrations are *perpendicular* to the direction of travel and are generally known as **transverse waves**. An interesting aspect of S-waves is that they can only travel through a solid, not a fluid.

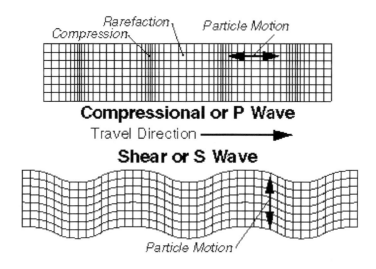

The differences in travel times, their unique shapes, and their ability to travel through different media make S and P-waves easy to identify and are handy data when studying the interior of the Earth. Seismologists set up multiple seismic wave detectors across the globe. When an earthquake happens, these detectors register the different times and types of waves that arrive and can pinpoint the exact location within the Earth that the quake had occurred (its hypocenter). Even more surprising is the fact that using this technology and science, we can visualize the layers of the Earth, their depths, and their physical properties due to these waves. We now know that the outer, thinner, layer of the Earth is the rocky crust, the next and thickest layer down is the mantle, the third layer down is the very hot, and liquid, outer core, and finally, there is the solid inner core.

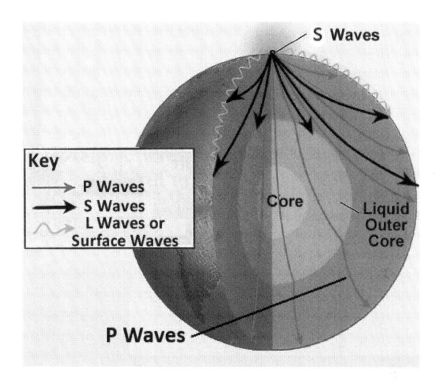

The S-waves cannot penetrate through the Earth's outer liquid core. So there happens to be a valuable phenomenon called an "S-wave shadow" on the other side of the planet when registering an earthquake. This also adds to the data points that seismologists use to determine locations of these events and the material properties of the interior of our planet.[15]

---

[15] There is also a "P-wave shadow" that occurs as well. But this is not due to a P-wave's inability to travel through the liquid outer core. Indeed, they do pass through. It is due, instead, to the refraction, or curving, of the P-waves through

This is but a taste of the fascinating study of seismology.

Properties of Waves

So far, we have been dealing with single waves at a time, but something peculiar happens when multiple waves come together and interact. When we picture a wave, let us just focus on the transverse variety, since it will be easier to visualize. However, in a sound wave passing through air or water, only the longitudinal variety occurs. Imagining a transverse wave will not diminish the idea, but instead clarifies the concept. So, the portion of a longitudinal wave that is compressed has a higher density, so we may translate this as the peak of a transverse wave. Likewise, the rarefied portion of the sound wave can be likened to the trough of a transverse wave. This means you can easily convert between longitudinal and transverse waves if you like.

Now, if you watch very closely to two rings of water wave that approach each other, you will see a great demonstration of their **interference patterns**. These rings can be created by starting with a still pool of water. Then two pebbles or rocks are dropped into the pool at the same time, but a distance apart from each other. Each stone will create a ring of concentric waves that spread out from the impact point. What happens when two waves collide with each other? It all depends on what point in the wave cycle the collision occurs at. When it occurs, you'll notice that some portions seem slightly higher, and some that are somewhat lower, than the peaks and the troughs of the single

---

the Earth and the ever-more increasing distances between each of these waves that occur as the distance away from the source is.

waves. There are also areas where there appear to be no wave undulations happening at all.

When the peak of one wave coincides with the height of another, there is an event called **constructive interference,** and a higher peak occurs here. The same happens when two troughs meet but in the opposite direction. Here the resultant trough is deeper than a single one by itself. In both cases, there is an additive, or constructive, interference. Another way of saying that two waves are coincidental in their actions is to say they are "in phase."

However, when two waves meet up in such a way as to overlap troughs with peaks, this sets up the appropriately named **destructive interference**. This occurs when the waves are "out of phase." When a peak and a trough coincide, you can expect that they will cancel each other out, and that is precisely what occurs. You would see no up and down vibration of the wave in this area, as there are opposing actions happening at the same time.

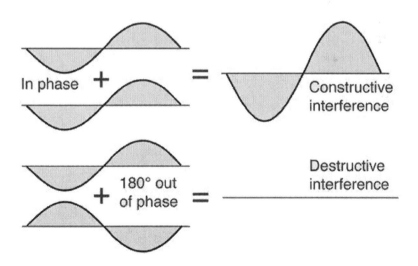

When multiple instruments are playing together in an orchestra, there are many different frequencies heard. Some of these sounds may overlap and constructively interfere, giving us a louder and more intense tone, while others will destructively interfere and provide either a lower intensity or completely cancel each other out. This symphony of sounds can get dramatically complicated.

You may have wondered how a sound can be heard from another room when there are so many walls and barriers along the way. Why don't they stop or absorb the sound? Sounds travel in three-dimensional shells (or bubbles) from their source. The edge of this expanding sphere is called the sound's **wavefront**. However, these shells expand in straight-line motion away from the source. If a wall happens to be in the way of the wavefront, what happens to the wave? There are several events that occur and can depend on the material of the wall. Some materials may absorb more of the sound waves than others. **Absorption** of sound occurs when the particles of the material being hit take up some of the vibrations of the sound from the medium and strip some of its energy away. Softer materials, such as foam, absorb sounds better than harder substances, like walls.

A memorable example of sound absorption comes to mind. Growing up as a child in Washington state, we used to get heavy snowfalls during some Winters. Going outside to play in the snow was always looked forward to with great anticipation. However, when going outside to play, I was always struck by an irritating curiosity that nagged my attention standing there in the snow. A strange sense of loneliness would overtake me. Everything around me always seemed so *quiet*. I could never put my finger on why this seemed to happen after it snowed. Indeed, it was disconcertingly too quiet. The reason, I later found out, was that snow is a rather good absorber of sound. Typically, the ground and the objects around you reflect a lot of the sound hitting them, but when snow abounds, sounds fade away. (Not to mention, that snow also tends to have that effect of slowing and shutting down noisy traffic as well.)

On the other hand, **reflection** off of a wall is what happens to most of the sound that impinges upon it, although there is a fraction that still gets absorbed. While the molecules of the wall may very temporarily absorb the vibration of sound from the incoming wavefront, it then releases this vibration back to the medium from which it came. This is the principle behind reflection. So, a single wavefront can get bounced and reflected from wall to wall, making its way to your ear around a corner.

There is also another interesting phenomenon, called **diffraction,** that occurs when a wave encounters the edge of a barrier (like a wall), bends around it, then continues forward. This is another reason why we can hear sounds around corners.

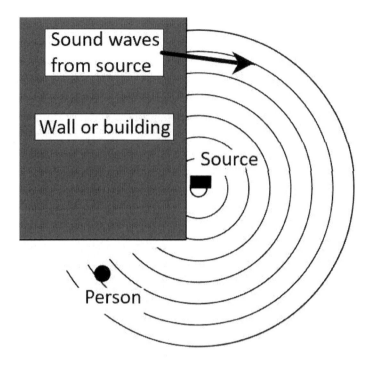

There is one more concept that applies to waves that we'll discuss, and that is the effect that passing through different media has on a sound wave. Remember that sound waves are waves of compression and rarefication of particles. Also, we had discussed how sound travels at different speeds through a medium depending on temperature. The higher the temperature, the faster the speed. Well, what happens if a sound goes through a medium that gradually changes temperature? Let us take the scenario where we have the air closer to the ground cooler than that found at a higher elevation. Additionally, direct a sound wave at an angle upwards. We will see that the sound will gradually curve downward toward the ground at an arc. This is because through the cooler air, the waves travel relatively slowly and as it passes through warmer and warmer temperatures, the waves pick up more and more speed. So, then why doesn't the wave just stay in a straight line going faster and faster at the original angle?

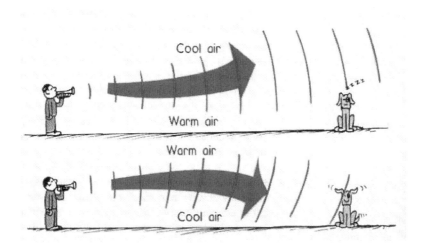

The trumpeter's sound gets refracted in an arc toward the cooler air. If the cooler air happens to be above the dog, which is farther away from the trumpeter, then the snoozing dog will not notice the sound.

The following explains why. The wave is broad, and the uppermost edge of that wave begins feeling the warmer temperature before the cooler lower edge. That higher, more heated, portion of the wave picks up the faster speed before the lower portion does and so there is a slight differential in rates across the entire width of the wave. Imagine two wheels connected together by a long axle. One wheel is on the left, and the other is on the right. They are moving forward. This double wheel represents our wavefront traveling forward through the air. As the left wheel of the wheel-axle system gradually crosses through a warmer part of the road, while the right wheel is still on the cooler road, that left wheel begins to move faster. What happens to the wheel system if the left wheel starts moving more quickly than the right side? Is it easier to see how the system will begin to turn in a rightward arc? This is precisely what happens

145

when sound passes through gradually changing temperatures. (The opposite directions of motion would occur if the scenario were reversed. If the ground air was warmer than higher up, the sound wave would curve in an upward arc instead.) A sound wave curves in such a way as to seem to avoid the warmer parts of a medium. This change in wave direction is called **refraction**.

Refraction also occurs when sound waves cross from a medium of one density to that of a different density. Waves travel at a faster speed in denser media. For example, when passing from air to water, a sound wave will be refracted at a steeper angle toward the higher density water.

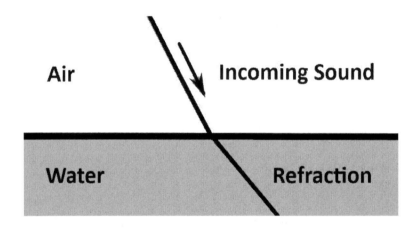

In this chapter, we studied periodic motions. We looked at how vibrational motions can be seen as rotations of a circle in a different plane. The behaviors of springs and pendulums were discussed and what periodicity means in harmonic motion. Periodic waves through various media were focused on, and how science and technology make use of sound waves through air, water, and ground.

Up to this point in our discussion of physics we've mainly focused on solid objects and their motions. Now, we will begin turning our attention to different states of matter. We'll see how they behave and what their properties are. Next stop: liquids.

CHAPTER 9
LIQUIDS

"Everybody is a genius. But if you judge a fish by its ability to climb a tree, it will live its whole life believing that it is stupid."

- Albert Einstein

Pressure

Earlier, when we discussed what forces were, we defined them merely as something that provides pushes or pulls. In ideal models, these forces are applied *to* a single point, such as an arrow point pushing on a point mass, or they emanate *from* a single location, like gravity from the center of the earth. However, these are overly simplified versions of the real world. There are no such things as dimensionless point particles, and there is no single point from which gravity comes from within the earth. The entire mass of the planet contributes to the forces of gravity. Every particle on and within the Earth contributes to the gravity you feel. When we move out of the highly efficient, but cartoonish, world of models and into the real world, we find a more complicated word. This is not to say that the earlier models were wrong to use. Indeed, they are not. They have their place in idealized models and work excellently in those cases. But we need to realize that those models will not work in all cases.

Imagine an object, such as a box, being pushed by a hand. Earlier, we could have said that the hand was an arrow that provides a *force* at a single *point* on the box. However, the hand actually contacts a certain amount of *area* of the box's side.

There is no point contact, despite what we used, or discussed, earlier in our simpler models. The force of the hand can be considered to be distributed around the whole area of contact. In other words, there may be 10 newtons of force in the arm pushing through the hand. So, there are 10 newtons of force distributed to the palm, the fingers, and the thumb. Thus, the area of the box that happens to be in contact with the hand has 10 newtons being applied to it. When you use a force against every point in an entire area, this is called a **pressure**. Pressure is measured in newtons per square area. In SI units, this would be newtons/meter^2 (N/m^2). 1 N/m^2 is also known as 1 Pascal (Pa). If you're not into metrics, but into Imperial units, then pressure may be measured in "psi," or pounds per square inch.

In this chapter, we will be discussing liquids and their physics. Liquids, like any other matter, has mass associated with it. As we all know, mass takes on weight when it happens to be within a gravity field. Weigh a kilogram of water on a scale, and you'll see that it measures about 2.2 pounds (lbs) or 9.8 newtons (N). Weigh *two* kilograms of water within the same container, and you'll measure about 19.6 newtons. Keep adding water and the higher the water is in the container, the more mass there is and, obviously, the more it will weigh.

Now comes an interesting and simple experiment assuming you have access to a swimming pool and know how to swim decently. Blow up a rubber balloon and tie it off. Submerge yourself and the balloon. Take note of the diameter of the balloon (this is probably best with a mask on). Now begin to sink lower and lower into the pool (if you can!) while holding the balloon. What do you notice? The balloon shrinks! And, yes, your ears are popping! Both of these observations are caused by exactly the same phenomenon: water pressure.

It turns out that the lower you go in the water, the more weight of water there is above you. Imagine that one-kilogram container of water resting on your head. Next, imagine the container with two, then three, kilograms of water resting on top of your head. As you increase the amount of water, the weight increases, and presses against the top of your head. This is similar to what is happening within the pool when you are submerged. However, not only is the water above you exerting weight on your head, but also over the entire surface of your submerged body, including the soles of your feet, and is known as **fluid pressure**.

That balloon you brought down with you (and the air within it) was also feeling the pressure of the water and was being squeezed together and shrank as a result. But why doesn't the balloon shrink only on the top side of it where the water is pushing down on it? Why does the entire volume of the balloon shrink instead? That is because the fluid pressure is being felt all around the balloon's surface – the top, the sides, the underside, every part of the surface area. Do you remember Newton's third law of motion? Here's a refresher: every action has an equal, but opposite, reaction. If there happens to be water weight pushing down, there is also the same pressure pushing *up* against that mass of water. So, everywhere you are within a mass of fluid, there is a pressure that is exerted in every direction. And that pressure increases the lower in the liquid you are.

Buoyancy

Sometime shortly after 265 BCE, a young man runs wet and naked through the streets of Syracuse, Sicily, exclaiming, "Eureka! Eureka!" (Greek for "I found it! I found it!"). This "Wet n' Wild" episode of antiquity was brought to you by physics.[16] What made Archimedes so excited as to forgo his dignity with the neighbors?

Archimedes happened to discover a relationship between buoyancy and weight within a fluid, and this principle bears his name – he just happened to be bathing in his bathtub at the time of the event. We know what weight is, but what is buoyancy? **Buoyancy** is a force like weight that acts *upwards* on an object in a fluid. In our previous section on pressure we discussed how water weight creates increasing pressure within deeper levels of

---

[16] The source of Archimedes' story is from the ancient Roman architect Vitruvius and his architectural masterwork *De architectura*, but may be an apocryphal one.

water. It is this weight of water, within the water itself, that creates the force of buoyancy. Let's explain this in more detail.

When an object is submerged within a fluid, that fluid must be pushed aside to allow enough volume to be occupied by the newly introduced object. For example, if a ball that has a volume of say, 10 cubic centimeters (10 cm^3), were pressed into a container of water it will take up 10 cm^3 of space that the water in this volume used to occupy. That 10 cm^3 of water must be pushed aside. There is a weight associated with that volume of moved-aside water. It is that weight of pushed aside water that creates the buoyancy and pushes upward against the object. The takeaway point is that *buoyancy = weight of the displaced fluid*. This is **Archimedes' Principle** and was the cause of the embarrassing Syracusan street incident.

Another way of imagining Archimedes' Principle is through the following. Take a container of water; a pool, say. A person is holding their breath and swimming within the pool. Now, imagine carving out a volume of water within the pool that is the shape of that person. This would be like a liquid water sculpture of the swimmer. Take that fluid sculpture and weigh it. The *weight* of that carved out volume of water will be equal to the *upward force* exerted upon the swimmer from the surrounding water. This upward force is called buoyancy.

Let us again take an object and place it on top of a fluid. Will it sink or float? It all depends on the fluid density and the weight and volume of the object. But, let's just stick with water for the moment. The object pushes down, due to its weight, and buoyancy pushes up on the object. These two competing forces are playing tug-of-war on the object and the one that is largest wins. If the weight of the object is greater than the buoyancy force, then the object will sink. If the buoyancy is the greater force, the object will float to the top of the water. And if they are both equal? The object will hover within the water and will neither sink nor float to the top.

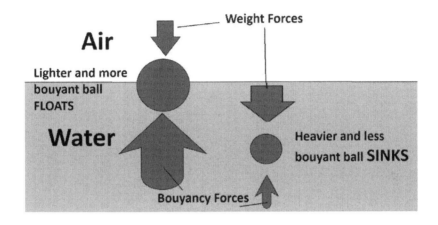

Something interesting happens within a fluid, though, as you sink through it. The buoyancy force becomes greater with depth. This is because of the more significant pressure of the liquid, the deeper within you are. So, let's say that the object that was dropped within the fluid is heavier than the buoyancy force, but only by a little. Obviously, the object starts to sink. But as it gets deeper within the liquid, it begins to feel an ever-increasing buoyant force acting upon it, until it reaches a depth where both the weight of the object and the buoyant force are equal. What happens to the object at this point? It stops dropping! It remains submerged and buoyant within the water.

Fish can stay at a static level within water due to an internal organ they have, called a swim bladder. This is essentially a gas bag that the fish can adjust to change their innate buoyancy within the water. If they choose to sink a little, they tighten the muscles around the bladder to squeeze and shrink its size. This creates a smaller fish volume and the fish sinks. To rise, the fish releases the pressure on the air bladder, and it expands, making the fish ascend. The air within the bladder can be renewed by gulping air at the water's surface. (Interestingly, the fish's air bladder has been thought to be the evolutionary precursor to air lungs in other animals that no longer live in water.)

It makes one wonder how on earth a gargantuan cruise ship can float on water instead of merely sinking to oblivion. It all boils down to the principles of buoyancy. Yes, ships can be gigantic metallic monstrosities, but sea-worthy they are. Much of a ship's volume consists of air. As of this writing, the largest cruise ship in the world is the Royal Caribbean's *Allure of the Seas*. It has a length of 1,188 ft (362 m) and weighs over 224 million pounds! However, the weight of the displaced water from the floating ship is much more substantial than this, and therefore, the buoyancy is higher, keeping the boat afloat.

"Eureka! Eureka!"

Let's go back to those streets of Syracuse where the citizens were surprised to see that strange Archimedes character screaming about something he had found in his tub. The story goes that he was given a job to determine whether the Syracusan Greek tyrant Hiero's gold crown was, in reality, made entirely of pure gold. Hiero gave gold to a goldsmith to form a crown for him. After receiving the crown, he tasked Archimedes with determining the authenticity of it, as he suspected the goldsmith of cheating him and adultering the crown by mixing the gold with bronze (and, thereby, stealing the leftover gold). One day, after cogitating on the matter for some time, Archimedes decided to take a bath. As he stepped into the tub, he noticed a curious thing.

When submerging himself within the water, the level rose around him. Archimedes' weight and the displaced water converged into that shout of "Eureka!" He now had the key to determining the authenticity of the allegedly gold crown.

This is how it worked. Every kind of material, including gold and bronze, has a specific and unique density associated with it. **Density** is simply a measure of the amount of mass within a given volume. So, density is in units of mass/volume. Gold is denser than bronze and Archimedes knew the differences. What Archimedes had, pre-eureka-moment, were gold's natural density and the crown's weight, but he did not know the crown's volume, as this was very difficult to arrive at, considering that the headpiece was oddly shaped. How could he possibly arrive at such an intricate value? Fast forward to post-eureka-moment, and Archimedes had his answer: submerge it in water and measure the volume of water displaced! He did this experimentally and reportedly found that the crown was indeed mixed with bronze as it turned out that the density of the crown was less than what would be expected if it were made entirely of gold.

What the eventual fate of the goldsmith was, is anybody's guess, but it was most likely not a good one!

Cohesion and Adhesion

Have you ever been in a situation where you've washed your hands and attempted to dry them, but have been frustrated with drying that dang water off? Usually, it arises when you're in a hurry, but why does that water stick to your hands so adamantly? Yes, water is sticky! Now that it's been mentioned, why does anything stick to anything else in the first place? Like Scotch tape or glue or cement, how does all this stuff work?

The physics of that sticky stuff is separated into two categories. One is called adhesion, while the other is called cohesion. **Adhesion** means the sticking together of *different* materials – wallpaper and wall; tape and paper; glue and wood; brick and mortar. **Cohesion**, on the other hand, means the binding of *similar* substances – water binding to water; liquid mercury binding to mercury, etc.

Let's talk about adhesion first. There are several phenomena that contribute to an adhesive's stickiness. We need to get down to microscopic and submicroscopic levels to appreciate these oddities. Cement mortar is a quintessential **physical adhesive**. If you look at it very carefully, you'll see that it consists of tiny pieces of rock or sand, water, and a thin paste. Bricks, at this level, contain innumerable pores, valleys, and rough edges. When slathered between two bricks, the mortar will slowly wet and fill the brick's pores and around all of their edges. Once the water evaporates and cures the cement, the paste within hardens and becomes a solid matrix in which the particles of sand reside and add strength. The ability of physical adhesives to bind comes from their wetting (surface flowing) and undercutting (locking) characteristics.

Another kind of adhesion comes from chemical means. **Chemical adhesion** comes from events that occur at a much smaller scale. At the atomic and molecular scales, there are bonds called ionic, hydrogen and covalent. These all relate to the attractions of electrons and positive charges and the resultant binding together of their respective ions, atoms, or molecules. Superglues are great examples of these kinds of bonds. The glue's molecules permanently fuse together with the surface's molecules.

But what about that sticky water on our hands? Is that a chemical or physical adhesion? It is neither. There is yet another variety of adhesion. This kind is called **dispersive adhesion** and results from slight atomic attractive forces between the two surfaces close to each other. There is an interesting effect at the atomic scale called **van der Waals' forces**. Named after the 19th-20th-century Dutch physicist Johannes Diderik van der Waals, these are feeble attractive forces. When electrons are orbiting within an atom, there are very brief times when the electrons happen to spend a little more time in a different area of the atom exposing the rest of the atom to an electron-free "hole." This hole happens to be slightly more *positive* due to the deficit of electrons in the area and the greater exposure of the positive nucleus of the atom. The positively charged area attracts the more negatively charged areas on the surface nearby and bonds them together very briefly. Over a period of time, these positively and negatively charged portions of surface areas interact and bond with one another.

There is a particular form of van der Waals force that applies to water molecules, called **hydrogen bonding**. Water molecules are shaped like Mickey Mouse heads. The smaller "ears" of the molecule are where the hydrogen atoms are, and the larger "head" is the oxygen atom. The oxygen atom of the water molecule pulls stronger on the hydrogen's electrons and keeps those electrons farther away from the "ears," leaving those areas more positively charged and the oxygen portion of the molecule more negatively charged as a result. These charges tend to line up with opposing charges nearby, including those found on the skin of hands. So, the negatively charged oxygen atoms within the water molecule tend to attract the positively charged hydrogen portions of other water molecules, and they weakly bond together forming those hydrogen bonds (see the illustration below). Well, these same kinds of negative and positive charge bonds can happen between the water and your hands as well. They attract each other and make the water a bit sticky.

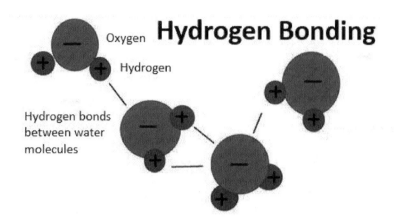

Have you ever noticed that tiny upward curve right at the water's edge where it makes contact with the surface of the inside of a cup? Look next time. This is called the **meniscus** and is caused by the same sticky interactions between water and hand. The water binds ever so slightly to the surface of the glass container. After washing dishes, you'll also notice that water droplets like to stick to everything! This is all due to those hydrogen bonds and van der Waals forces.

There are other less common forms of adhesion as well, but we will not detail these here.

The other broad category of bonding is cohesion. For example, using water again, it likes to stick together. Slowly turn on the tap and notice how water falls in a collective cylinder rather than a wide spray. Or watch on a rainy day as the droplets of water hit your glass window, slowly fall down, but then seem to have affinities for the nearby water droplets, and subsequently merge with them. That is water cohesion at work. In this case, likes attract. As mentioned earlier, water molecules have a peculiar structure that makes them a bit "polar" in the jargon of chemistry. This means there is a difference in charge across the molecule – one end has slight negativity, and the other has a mild positivity.

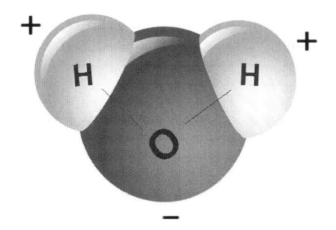

The polarity of water molecule

So, in liquid water, all these molecules are moving around, but they like to hang together slightly in chains where the positive ends seek the negative ends and vice versa. Macroscopically (at our scale), this appears as water sticking together. This cohesion also gives small, free-standing liquid water its spherical shape. If the droplets become too large, they start to flatten out from their increasing weight.

Remember the water meniscus, where water likes to have an upward curved lip on the surface that it touches? Well, put liquid mercury (a heavy metal) within a container, and just the opposite happens. Oh, it forms a meniscus, but it's a *downward* curved meniscus. This is because mercury forms a stronger cohesive bond and a weaker adhesive one. Mercury also forms droplets on flat surfaces, but these droplets can be larger and retain their sphericity better than water droplets do.

The concave meniscus of water and the convex meniscus of mercury

Viscosity

Ah, the frustrations of some of the simpler physics in life. Despite all of the shaking, tapping, tipping, and yelling at the newly opened and upside-down ketchup bottle, the fries are still awaiting their immersion in crimson goodness. Better yet, attempt to get out that last bit of peanut butter from the plastic jar! Who was it that invented viscosity? They should lose their job!

160

However, I come back to reality from my daydreaming and start appreciating the reasons for these viscosities. Every gaseous and liquid substance has some sort of viscosity associated with it – some more than others. **Viscosity** is simply how flowable a fluid is or how much resistance to movement it has. Water has much less viscosity than honey. While honey has much less viscosity than peanut butter. However, a substance's viscosity can also be affected by temperature. Warm up that peanut butter, while keeping the honey in the refrigerator and their respective viscosities may reverse.

The physical explanation for viscosity is similar to cohesion. The molecules from which substances are composed, come in an infinite variety of shapes and sizes. Within each of these molecules, there may be spots that have more accumulated electron charges than at other locations within the same molecule. In addition to this, some molecules may be long chains that can have multiple branches or portions that stick out laterally. When the molecules slide by one another, they have a kind of molecular friction that gives them a resistance to fast movement. This is caused by the bumping up against these branches, the electrical attraction at some charged points on the molecules, or a combination of both reasons. All of this accumulated resistance to movement contributes to a fluid's viscosity.

One of ketchup's long-chained molecules

While edible viscosity can pose a frustrating problem for some, they can make up for it by looking at some of the amusing sides of thick viscosity with a little adhesion mixed in. My favorite demonstration of these two combined qualities is by scooping up some goopy peanut butter on a spoon and applying it to the roof of my dog's mouth. The more cynical among us may say that I've simply moved the frustration from one beast to another. I must demur, and state the fact that my little dog gets the added benefit of exercise and delicious nutrition at the same time, all while giving those who watch a great time!

Bernoulli's Principle

The last concept we will deal with in this chapter is associated with all fluids, not just liquids. It is called **Bernoulli's Principle**, named after the Swiss mathematician and physicist, Daniel Bernoulli (1700-1729). Bernoulli's Principle states:

"A fluid that flows with increased speed does so with a decreased pressure, and vice versa."

This principle of fluids is so important that it is what keeps us up in the air when flying in an airplane. Less importantly, it is the principle by which perfume misters, liquid soap dispensers, hose nozzles, car carburetors, so-called pitot tubes on aircraft for measuring airspeed, and many other applications use. However, to understand this counterintuitive concept, we'll need to work our way up to it, starting from more straightforward principles of fluid flow.

Let us picture a fluid, like water, flowing through a simple round cylindrical tube. The water flows from left to right. Now, think of an imaginary round cross-sectional area in the tube. We'll call this area A1. As water flows, it will pass through this imaginary round wall within the tube. We know that as much water that comes from the left side of the wall must be precisely the same amount of water that crosses to the right side of the wall, right? Correct, and importantly, *the velocity of the water will be the same on the left and right of the cross-section, as well.*

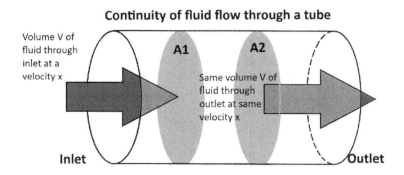

Fluid flow continuity

Let us add another imaginary wall farther to the right of the inside of the tube from where A1 is. We'll call this new cross-sectional area A2 (as shown in the above illustration). It is also simple to see that the same rate of water flow that crosses A1 must also pass A2. This means that the amount of volume of water that crosses A1 every second is the same as A2. For example, we can say that 10 gallons of water every second crosses A1 and A2. This must be so because imagine what would happen if this weren't true. Imagine 10 gallons per second crossing A1, but only 5 gallons of water per second crossing A2.

There would be a bunching up of water between the two, but this is physically impossible. Water is incompressible, so it is impossible to squeeze more water into a specific filled volume than what is already there. So, if 10 gallons per seconds crosses A1, we know that 10 gallons also passes A2 at the same time. Again, *the velocity of the water flowing across area A2 will also be the same as that crossing area A2.*

So far, we have been dealing with a simple length of tube that is the same width (cross-section) all the way through. We need to raise the level of complexity a little to get to our final important fluid concept. We need to make a change in the shape of the tube. On the left half of the tube, it will be twice as wide across as the right half of the tube. So, there is now a constriction in the tubing. Fill the entire tube with our water again. Apply pressure to the water from the left side. How will the water flow in the tube now? According to Bernoulli's principle, the water on the broader left half will flow at a slower velocity and with a higher pressure than the flow on the right half.

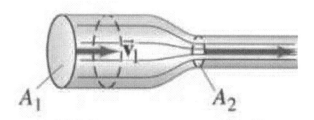

Bernoulli's Principle. The slower water velocity, v_1, passes through the wider cross-sectional area, A_1, of the tube. The faster-flowing water crosses the smaller cross-sectional area A_2.

The way we explain this is to first understand what was mentioned earlier – that the same volume of water per second crosses every cross-sectional area within any part of the tube, large or small. So, place two imaginary cross-sectional areas, A1 and A2, within this new tube. A1 within the wider section and A2 within the thinner section. In this case, how can we explain the fact that the same volume of water per second crosses A1 as A2 if one area is larger than the other? The answer is that as a specific volume of water per second passes through A1, there needs to be a *smaller volume moving at a faster rate* crossing area A2.

The counterintuitive and challenging concept of Bernoulli's principle comes from the idea that the *faster*-moving water has *less* pressure. (Or, inversely and just as strange, the *slower* moving water has *more* pressure.) We would typically suspect that faster-moving water seems to have a *higher* pressure, right? For example, place your hand under the open tap and compare the difference in pressure when the water flows slowly as opposed to quickly. Your hand will feel a higher pressure associated with the faster-flowing water.

However, this is not the pressure that Bernoulli's principle is related to. The principle is concerned with the *inside* fluid pressure within the tube, not the pressure of water coming out of the tube. These happen to be two different scenarios and, as a result, two different pressures.

Let's look at the best classical example of Bernoulli's principle in action. This happens to be the lift that the wings of aircraft experience. The side view shape of a wing is called an **airfoil**. An airfoil looks like an elongated teardrop shape with a round front end and a tapered and thin back end. The underside of an airfoil is flat, while the top is curved. When an aircraft travels forward through the air, the airfoil cuts the air in front of it in half, and the two halves of the airflow travel along the upper and lower body of the wing. The air flowing on the underside of the wing travels alongside it slower than the flow above it.

Because Bernoulli's principle says that the slower the airflow, the higher the pressure, there will be a pressure difference set up between the top and the bottom of the wing.

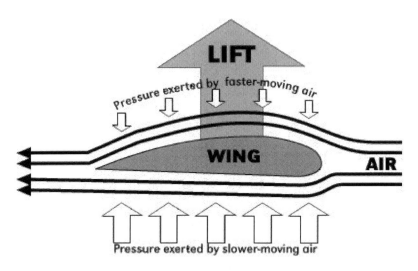

Lift of an airfoil (airplane wing) caused by higher pressure on the underside and lower pressure on the upper side.

Higher pressure air is found under the wing, while lower pressure is on the bottom. The higher pressure will push up on the wing. The lower pressure upper side of the wing is overwhelmed by the underside's higher pressure. These actions create an upward force on the wing called **lift**. Lift is the reason that planes stay in the air when moving forward. The next time you disembark from an airplane, remember to always give thanks to physics in your prayers, for getting you from point A to point B in your fast, safe, and reliable travel!

This chapter dealt with some of the physics of liquids. We swam underwater to demonstrate pressures and buoyancy. We traveled to the ancient world to discover the equivalency between weight and displaced volume. We almost threw a newly opened ketchup bottle across the room, but instead, we came to our senses and became patiently appreciative for our study of the physics of viscosity. Finally, we looked at the counterintuitive ideas behind Bernoulli's principle and gave thanks to physics for saving our lives, once again.

On our journey through ever more thinner states of matter, we will next make a stop at the gossamer world of gases.

CHAPTER 10

GASES

"Love is in the air? Wrong. Nitrogen, oxygen, argon, and carbon dioxide are in the air."

- **Sheldon Cooper**, *The Big Bang Theory*

Density

Take a look at the periodic table of elements. This chart shows, in a very orderly way, a listing of all the chemical elements that are found in the universe, including some that are not naturally found but are instead man-made (those that are beyond Uranium, atomic number 92). At our ordinary room temperatures, every element exists in a particular physical state – solids, liquids, and gases. If we restrict our attention to all those elements between, and including, atomic numbers 1 – hydrogen - and number 92 – uranium, we find that there are 2 liquid (blue in the diagram below), 11 gas (yellow) and the rest are solids (green).[17]

---

[17] Those elements that are liquid at room temperature are bromine and mercury, while those that are gases are: hydrogen, helium, nitrogen, oxygen, fluorine, chlorine, neon, argon, krypton, xenon, and radon.

1 H																	2 He

Periodic table showing elements with states of matter key (Solid, Liquid, Gas).

Assuming the right temperatures and pressures, all these elements can be converted to any one of these three states of matter. So, what exactly makes a particular element a solid, a liquid, or a gas? It depends on that element's average particle spacing, and that can be measured with density. **Density** measures how much matter exists within a given volume of space. Get a jar and fill it one third with sand. On top of the sand add water to fill the next third of the jar. You now have a jar that contains three states of matter within it. The sand layer is solid (albeit with lots of air spaces between the grains)[18]. The middle third consists of liquid water, and the top third is gaseous air. Each layer occupies the same volume – one-third of the entire jar's space. The sand, being a solid, has more mass per unit of volume within it than the water or the air. The liquid water has less mass per volume than the sand, but more than the air and the

---

[18] Yes, in reality, the air gaps between the sand grains will probably fill in with water as well, but we can ignore this for now and ideally picture the sand as simply a solid layer.

169

air has less mass per volume than the other two layers. In order of highest to lowest density, then, we have the sand, the water and then the air, respectively. This density difference is also what keeps the layers separate from each other and in the order that they are found.

Intuitively, you may think that *all* substances exhibit this higher to lower density order from solid, to liquid, to gas, but this isn't always the case. It is a good rule of thumb, but a noteworthy exception to the rule is water. Its order of density from highest to lowest is *liquid*, solid, gas. This is the reason ice floats on top of liquid water; it happens to be less dense. Incidentally, this property of water is very fortunate for fish who must swim and survive within this medium. Since standing water freezes beginning from the top to the bottom, this leaves the rest of the water volume, down at the bottom of the depths where food may gather, unhindered by ice. This allows fish to go about their fishy business without the risk of freezing within a solid block of ice Han Solo-style.[19]

Gas Pressure

How did we establish that the air around us takes up any space at all and that it is composed of a physical substance? This idea was first demonstrated by the ancient Sicilian Greek philosopher Empedocles (495 – 435 BCE). With the use of an ancient instrument called a "water thief" or clepsydra, Empedocles showed that air takes up space and has a specific

---

[19] Arctic and Antarctic fish have an additional life-support backup system in the form of a special kind of protein antifreeze within their blood. With this protein, they are able to withstand the freezing temperatures and still retain a liquid state of blood within their bodies.

volume. The clepsydra was either constructed out of glass or brass and had the shape of a bulb that tapered into a tube at the top with an opening. At the bottom of the clepsydra were multiple small holes.

A bronze clepsydra, or "water thief."

This instrument could be held by the neck and submerged in water to fill it. Once raised out of the water, the bottom holes would spill out the container's water. It could be used to sprinkle plants with water. However, if you submerge the clepsydra while holding your thumb over the top hole in the neck, no water would enter the instrument. Release your thumb and water will enter. If you fill the bulb with water and then hold your thumb over the tube's hole, the water will not drop out. How did Empedocles explain this?

We know in our present-day that air does take up space, but in ancient times this was not so obvious. So, in Empedocles' time, there was a possibility that within the clepsydra, there existed water and a vacuum (void). The reasoning that Empedocles used for the operation of this device was that if a thumb closed the neck's hole and no water came out, this was because, for the water to escape, there would have needed to be something that would enter to take the space of the emptying water. Since the thumb closed this possibility, the water could not flow out. Once the thumb was released, that "something" could now enter from the top of the clepsydra through the hole in the neck to replace the continuously missing water that dropped out. Likewise, if a thumb were closed over the neck's vent when the clepsydra was dipped into water, no water would enter the device, since there was no escape passage available for that "something" within to be replaced by the entering water. The mysterious "something," Empedocles reasoned, was air itself.

So, the air was given a substantiality and a real volume associated with it. Empedocles was historically the first to propose that all of nature was composed of four elemental "rhizomata," or "roots." These were air, water, earth, and fire.

It took the 17th-century Italian physicist and mathematician Evangelista Torricelli (1608 - 1647) to prove that air has weight. His experiment to show this was with the use of his invention, the mercury barometer. Torricelli filled a long glass tube with mercury, placed his thumb over the open end, inverted it upside down and gently submerged the open end within a bowl of mercury. When he released his thumb, the mercury within the tube lowered, and that within the container raised just slightly. The empty space within the top of the glass tube that had been filled with mercury now became a small vacuum (called a "Torricellian vacuum"). The height of the column of mercury in the tube was found to be about 760 mm (29.92 in).[20]

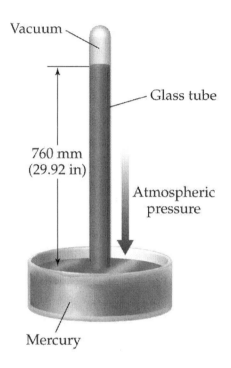

Vacuum

Glass tube

760 mm
(29.92 in)

Atmospheric
pressure

Mercury

What is occurring with the Torricellian barometer is a pressure that is being applied to the surface of the mercury exposed to the air within the bowl. This pressure pushes on the mercury, and that pressure gets transferred throughout the liquid (remember our discussion of this in the last chapter?). That pressure pushes the liquid mercury up the glass column until the

[20] Torricelli decided to use the heavy element mercury, rather than water, because mercury weights more and, as a result, will allow a shorter glass tube to be used. If water were to be used within the tube, due to its lighter weight the column's height would then be 10.34 meters (33.9 feet) high! Mercury is 13.6 times denser than water and will therefore make a column 13.6 higher than mercury.

weight of that cylinder of mercury exactly equals that of the atmospheric air pressure outside the bowl. Torricelli reasoned that his barometer actually measured the *entire weight* of the atmospheric air above his barometer all the way up to space! How was he to prove this hypothesis? He decided to test it by allowing the surrounding atmospheric pressure to be reduced. Well, this can be accomplished by moving to higher elevations. The higher up you go, the less atmosphere will be above you and, consequently, the less pressure weighing down upon you. Torricelli took measurements in his lab and then took his apparatus to the heights of a mountain and took measurements up there as well. Lo and behold, his hypothesis was supported! The pressure *was* reduced on the heights of the mountain, and the height of his mercury column was less. Air had been weighed.

Remember, that pressure is the measure of force upon an area. So, air pressure is gas pressure. But although pressure is a distributed force along a surface area, *it is considered a scalar, and not a vector*. It only has a value, but not a direction. It is assumed that whatever directionality that may be associated with pressure is always perpendicular to the surface area it presses against.

In 1648, the French physicist and mathematician Blaise Pascal (1623 - 1662), coined the term "pressure" to define the weight of the air in the Earth's atmosphere. The SI unit of pressure, the Pascal (Pa), is named after him.

You may be asking what the weight of gas has to do with that gas' pressure. Weight of a fluid is what creates the pressure of the fluid from within. Every action has an equal and opposite reaction. So, when we discussed pressure and buoyancy in chapter 9, we saw that the weight of a column of water above a certain level pushing down on that level has the reactive effect of an equal, but opposing, upward pressure being felt on the water above. *This pressure is distributed in all directions, though.*

There is just as much pressure being directed to the left as to the right as there is upward and as there is downward. Water is a fluid, just as gases are. They both behave the same way concerning pressures. Sink deeper into a body of water, and you'll feel your ears pop from the increasing water pressure on them. Do the same, but in the opposite direction, and rise higher and higher in an airplane. The air pressure begins to decrease, and again, your ears will pop as a result. None of these pressure changes depend on the orientation of your head, because the fluid pressures of the water and the air are pointing in all directions from within the fluid. Fill a balloon with gas, and you'll see this to be true. The balloon will expand symmetrically in all directions.

Boyle's Law

Going back to the inflated balloon that we took with us underwater in the last chapter, let's examine the balloon's properties again. When we (and the balloon) sank deeper into the water, we noticed the balloon shrank in size. When we ascended again, just the opposite happened – the balloon expanded. This is obviously due to the surrounding pressure on the balloon. The water pressure increases with depth and presses on all sides of the balloon, and its volume shrinks. Essentially, the air within the balloon is being squeezed to a smaller volume. This is a demonstration of **Boyle's Law**. This law is named after the English-Irish chemist Robert Boyle (1627 - 1691), whose experiments on gas showed the relationships between volumes, pressures, and temperatures.

Boyle's Law states that if you keep the temperature of a gaseous system the same, the pressure on the gas and its volume occupied are inversely proportional to each other. Increase one, and you decrease the other and vice versa. Another way of stating this is if you multiply the pressure and volume of a gas together that result will always remain constant. So, no matter what volume and pressure there are, the multiple will always stay the same constant number.

Boyle's Law: $P_1V_1 = P_2V_2$

Decreasing volume increases collisions and increases pressure.

$V_1 = 1.0 L$
$P_1 = 100$ mm Hg

$V_2 = 0.5 L$
$P_2 = 200$ mm Hg

The explanation for Boyle's law comes from what happens at the submicroscopic scale. Let's shrink down to see what is happening to those gas atoms and molecules. Gases are composed of tiny particles that are either atoms or molecules (composites of multiple atoms bonded together). These individual particles are flying around haphazardly at very high speeds. The speeds are dependent on the temperature of the substance. Occasionally, the particles will collide with either themselves or with the walls of the balloon (or whatever container walls there are). The force of the collision pushes on the wall, particle, or balloon surface. Billions of these collisions are happening every second within a balloon and keep the balloon's elastic material stretched continuously and retains a particular volume for the enclosed gas. Adding up all these infinitesimally small particle forces together along the entire surface amounts to the total pressure of the gas within.

If this gas must expand to a larger volume, what happens to the pressure of the gas within? It decreases because there will be more surface area of the container to hit, but the same number of particles going the same speed as before. This is equivalent to the same force being distributed to a larger area; that is, a smaller pressure. Let's reverse the scenario: shrink the volume. Smaller volume, smaller area to hit, the same number of particles, same forces. Well, a smaller area being pushed with the same force amounts to a *larger* pressure. So, lower volume and higher pressure.

Depth	ATM	Air Volume	
0 m	1	1	
10m	2	1/2	
20m	3	1/3	
30m	4	1/4	
40m	5	1/5	

In the world of SCUBA divers, Boyle's law can be a real problem. The blood within all of our blood vessels contains dissolved gases, such as oxygen, carbon dioxide, and nitrogen. The oxygen is tied up and attached to the red blood vessel's hemoglobin (this is what makes blood red). The carbon dioxide is dissolved in the form of bicarbonate ions. However, nitrogen is inert and is dissolved in the blood as tiny bubbles. Think of them as microscopic balloons within the blood. This nitrogen is continuously being released from the blood through our exhalations.

However, something hazardous happens if a SCUBA diver, deep underwater, rises too fast through the water. The nitrogen within their blood does not have time to be released and exhaled and instead expands due to Boyle's law. When the diver rises through the water, the surrounding water pressure decreases and, as we know, gases expand in this condition. Therefore, the dissolved nitrogen bubbles expand within the blood vessels and block off the flow of blood within. This creates a condition called "the bends" and is an excruciatingly painful condition to all of the muscles and joints. Divers have died from this "decompression sickness." The treatment for it is to immediately move the diver to a decompression chamber. This is a room whose air pressure can be adjusted. Once the diver is placed in the chamber, the air pressure must be increased to shrink those nitrogen bubbles in their blood. Then, the air pressure within the chamber is slowly decreased back to normal atmospheric pressure while the diver exhales the accumulated nitrogen.

The rarified state of matter that we call gas was discussed in this chapter in all of its glory and physics. We saw how density helps define what determines the different states of matter. The ancient philosopher Empedocles was introduced in our discussion of how the air was "discovered" and how he gave it an existence of its own. Weights and air pressures were discussed, while Mr.

Boyle's Law was looked at, and it was shown that a gas' volume and pressures are very much related.

The discussions of gases dovetail well into the next analysis we'll make. The study of thermodynamics is where our next focus will be in the next chapter.

CHAPTER 11

TEMPERATURE

"When you can't make them see the light, let them feel the heat."

- Ronald Reagan

Hot and Cold

Go outside on a sunny Summer day and feel the warmth of the Sun on your skin. Go back indoors to the kitchen and open the freezer and feel the coolness from within. You are obviously experiencing temperature differences, but what is the nature of temperature? Are there actual material substances called "heat" or "cold" that is touching your skin?

The historical development of the ideas of heat and the instruments to measure it is a long and convoluted one. The notions of heat span from the 6th century BC Greek philosopher Heraclitus, who proposed that the universe was composed of three elements: fire, water, and earth[21]. The idea of heat then passes through another hypothetical 17th century AD element within combustibles called "phlogiston," then to the 18th-century fluid-like "caloric," and finally to the modern physics of thermodynamics.

There was one principle among these concepts that tied them all together: the idea that heat was what caused objects to become hot, and the *absence* of it is what makes them cool.

---

[21] The element of air was not going to be added to the list until Empedocles in the 5th century comes on the scene (see chapter 10).

Temperature is the term that is used to express the measured quantity of hot or cold.

The instrument that we are all aware of for measuring temperature is the **thermometer**. Thermometers are designed in various ways and work on different principles. Some are glass cylinders filled with a fluid that expands and shrinks depending on their heating properties. This allows them to rise or fall, respectively, within the glass tube. A background of regularly spaced tick marks gives the reading for temperature. Another design makes use of the shrinking or expansion of metals. The terminology for this size change property of metals with heat can be a mouthful and is called their "coefficient of linear thermal expansion." Every metal and substance have a unique coefficient of expansion. The thermometer design that uses this principle is called a bimetallic thermometer, and this is how it works. Take a very thin long strip of metal that has a relatively large coefficient of expansion (meaning, it changes shape relatively significantly with changes in temperature) and take a second strip made of a different metal or material with a smaller coefficient of expansion. Like sticking two pieces of paper together, bond the two strips together to form one long strip. Roll the strip up into a spiral and attach a pointer to the middle of the spiral.

Bimetal Strip
Two Metals Bonded Together with Different Coefficients of Expansion

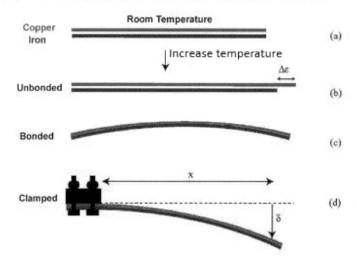

Bimetallic strips. (a) Take two thin strips of metal. One with a higher coefficient of thermal expansion (copper, in this case) than the other (iron). (b) Heating both will show a difference in expansion. (c) Bond the two strips together. (d) When one end is fixed, and the bimetallic strip is heated, the strip will bend in the direction of the lower coefficient metal.

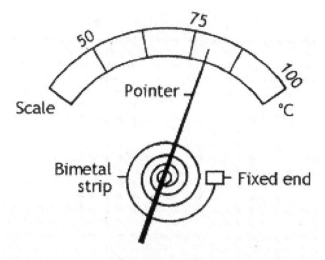

Bimetal Thermometer

The bimetallic strip is rolled together and attached at one end to a pointer, while the other is fixed. When the surrounding temperature increases the rolled strip expands and winds up tighter, moving the pointer to the right.

Have a calibrated background temperature scale behind the pointer and, voila, you have a bimetallic thermometer. When the temperature around the thermometer heats up, the larger coefficient strip expands more than the other piece and slides past it, causing the spiral to turn the pointer. When it cools, just the opposite happens, and the spiral shrinks and the pointer turns the other way.

Temperature Scales

There have been several scales that have been developed for the use of measuring temperatures. Three of the most common scales are the Celsius (formerly known as "Centigrade"), the Fahrenheit, and the Kelvin scales.

When developing scales, there is some kind of reference that is used in choosing the beginning point, the endpoint, or the size of the individual units within that scale. These kinds of temperature scales are called **relative temperature scales**. The other category of temperature scale is the **absolute temperature scale**, which we will soon discuss. Two of the most commonly used relative scales are the Celsius and Fahrenheit scales. In 1742, the physicist Daniel Gabriel **Fahrenheit** (1686 - 1736) proposed a system of scaling for temperatures. This scale is still used, but only in the United States, its territories, and some islands. Fahrenheit suggested the zero point of his scale at the temperature at which an equal mixture of ice, water, and a salt, froze. This point was called zero degrees Fahrenheit or 0°F. The freezing point of pure water, on this scale, is 32°F and the boiling temperature is 212°F (all at 1 atmosphere of pressure)[22]. In the Fahrenheit scale, this makes the difference between the freezing and boiling points of water 180 degrees apart.

---

[22] We must always give the surrounding air pressure when giving temperatures of reference. This is because if the air pressure is different, then the reference temperature can be different as well. Try boiling water at 212°F at sea level and attempt to do the same at a higher elevation, say, in Denver, Colorado. At sea-level, the surrounding air pressure is higher and is at what is called 1 atmosphere. However, at an elevation, the air pressure is *lower* than 1 atmosphere. At a lower air pressure, the boiling point of water will also be lower. So that water in Denver, Colorado will start boiling 10 degrees lower, at 202°F! It is important to be accurate when discussing temperatures and pressures. If no surrounding air pressure is given in the text, we can safely assume we are speaking of 1 atmosphere pressure.

The Fahrenheit temperature scale is a pretty arbitrary and odd scaling and one that doesn't lend itself well to scientific convenience. A better system was developed by the Swedish astronomer Anders Celsius (1701 - 1744) and one that is used as the standard in the rest of the world. The **Celsius** scale conveniently defines zero degrees Celsius (0°C) as the freezing point of water, whereas 100°C is recognized as the boiling point (again, only at 1 atm of pressure, as air pressure can make a difference at which temperature a substance freezes, melts or boils). Between the two extremes, there are 100 units, which makes scientific and mathematical usage much easier. Of course, in both of these scales, they extend indefinitely both higher than the boiling point of water, and lower than the freezing point (in which case, there can be negative degrees Fahrenheit or Celsius).

There is another scale that we will discuss, which happens to be an even better one to use in the field of science. This is called the **Kelvin** scale, named after the Scots-Irish scientist and engineer Lord Kelvin (1824 - 1907).[23] The Kelvin scale is an absolute temperature scale. This means that this scale is based upon a universal principle and is independent of any outside variables, such as pressure. Whereas the relative scales do dependent on these external variables, the absolute scale does not. Lord Kelvin calculated that there exists an "infinite cold," a temperature at which nothing else can be lowered. This point is called **absolute zero**. This is the temperature at which no more subatomic motion or vibration of particles occurs. This temperature, he calculated, was found to be -273°C. In the Kelvin scale, this is the zero point: 0 Kelvin (0 K). The more current and exact temperature at which no more movement of particles occur is -273.16°C. This means that water's freezing point is defined as 273.16 K. In the Kelvin scale, water's boiling point is at 373.16

---

[23] William Thomson, 1st Baron Kelvin, so-named after the River Kelvin that runs in front of his laboratory at the University of Glasgow, Scotland.

K. The units in the Kelvin scale are precisely the same size as that found in the Celsius scale. So, for every difference of 1°C, there is the same 1 K difference. Not so for the Fahrenheit scale, though. For every 1°C (or 1 K) change, there is a 1.8°F difference.

Expansion

Matter is made up of particles that are either vibrating, traveling on a path, or performing both motions at the same time. These particles can absorb or release energy in the form of electromagnetic waves, like light, infrared, ultraviolet, etc. When energy is absorbed by a particle, it gives a boost to that particle's motion, whether it is in a faster vibration, a faster traveling speed, or both. The reverse happens as well - when a particle releases energy in the form of electromagnetic waves, its motion slows down.

This is the interplay of matter and energy at the atomic scale, but what do we observe at the macroscopic scale of eyes and scientific sensors? When a material is heated, we generally see a physical **expansion** occur. When cooling takes place, the opposite happens, and the material shrinks. The vibration or speed of the particles, caused by the absorption of heat, increases their average distance from one another. If the material cools, energy is let go, and the individual particle's average distance of separation is decreased, and the matter shrinks.

When the substance expands, it does so in all directions, volumetrically. Take a block of metal and heat it up. The volume increases. This occurs in all the phases of matter: solids, liquids and gases. Long expanses of metal railroad tracks can heat up in warm weather or in direct sunlight and cause some significant buckling of the tracks from the ever-expanding length of the heated tracks (see picture below). If not compensated for, this has been known to be a major contributor of railroad derailments.

This concludes our brief discussion of temperatures and their measures. We talked briefly about the evolution of the concept of heat. The three major temperature scales were illuminated, as were the methods to measure them. Finally, we detailed what physical properties of matter change when heated. These were all pre-requisites for our understanding of our next, and last, topic of discussion in this volume: the all-important science of thermodynamics.

CHAPTER 12
THERMODYNAMICS

"Use 'entropy' and you can never lose a debate, von Neumann told Shannon - because no one really knows what 'entropy' is."

- William Poundstone

The Kinetic Theory of Gases

So, we've reached the chapter where we explain the nature of heat in greater detail. There is a branch of physics that deals with heat and all of its multiple interactions. This field of physics is called **thermodynamics** (Greek word origin from "thermis," meaning "heat," and "dynamis," meaning "power" or "energy").

When I was in college, majoring in physics, I found that I had loved every aspect of the science and consumed its teachings voraciously, and at every level, with one, rather embarrassing, exception. That exception happened to be the field of thermodynamics! To me, it always seemed to be almost a footnote or a side issue in physics. I do not know if this stemmed from my lack of reading much on the subject, the method it was taught to me, the arcane way it needs to be seen, or from some other reason. I did understand that the field was very pervasive in nature, but I was impatient to get on with my studies to begin the more exciting branches of physics, such as relativity theory, quantum mechanics, and astrophysics. In my impatience, I failed to realize how essential thermodynamics really was. I also was unable to see that much of our current and modern branches of physics, the physics I was so eager to learn, would not exist if it were not for the science of thermodynamics. This is the branch that underpins the very fundamentals of modern physics. Thermodynamics explains so much about our universe. In fact, its very origin and its death are tied to this science. So, I have come to recognize its true beauty and power, and now have a much greater appreciation for it.

Our first step in our pursuit of understanding will be to describe matter's behavior at the very small, the simple, and the diffuse. When we discussed gases earlier, we found that, sub-microscopically, they consist of innumerable atoms or molecules relatively far apart from each other and all speeding away in different directions. These particles continuously collide with either the container walls or with each other. This is the picture developed within the **kinetic theory of gases**. It is a useful model in science and helps simplify our analysis of gases. This means that not only can we visualize what is physically happening within a gas at very small scales, but it also has the added feature of *predictability* – a highly prized property of any theory in science.

But as with any model of nature, there needs to be oversimplification and assumptions. There are several assumptions in the kinetic theory of gases. Some are reasonable; some not so much. Here are a few:

- There must be an extremely large number of particles
- Each of the particles must be spherical and of the same size
- The particles are constantly colliding with each other or the walls of the container
- Particles do not interact with one another, other than from the collisions themselves
- Each collision (with walls or other particles) happens in what is called a "perfectly elastic" collision. This amounts

to a perfect bounce, where absolutely no energy gets left behind or escapes, such as in the form of noise or heat.

In the real world, the first point is very reasonable, as there can easily be trillions of particles within a volume of gas. The reason that this becomes important, is so that the gas can be treated and analyzed mathematically, through a branch of mathematical physics called statistical mechanics.

The second point concerning the sphericity and same size of every particle, however, is not very reasonable, unless we are dealing with pure elemental gases, such as helium. This is because in the real world, gases are rarely in a pure form and many times are mixed not just with other kinds of pure elemental gases, but can also be combined with more complicated molecular gases, such as carbon dioxide. Molecular gases are composed of particles called molecules, where each particle is a bonded collection of different atoms. Each of these particles may be a different size and shape. So, they may not be spherical nor of the same size.

The third point concerning collisions with walls and each other is much more reasonable as this is, indeed, occurring innumerable times within a given period of time.

The fourth assumption is that each collision is the only interaction that occurs between particles is not very reasonable, as there can be other properties that are felt between particles within the same volume of gas. These can include van der Waals' forces, hydrogen bonding, and additional electrical charges between particles, along with magnetic interactions.

Finally, the last assumption mentioned is an overly simplified brush-over of reality. Atoms may bounce off each other in perfectly elastic collisions, but *molecules* hitting each other do not. Particles that collide with the walls of a container do not do so in a perfectly elastic way either. Heat really is released and absorbed as is the energy diffusion in noise, etc. So, in reality, most of these collisions are of the *inelastic* variety and *do* lose energy to heat, noise, etc.

One of the interesting things about the kinetic theory of gases is that it can also be applied to liquids. All of the assumptions made for gases may be applied to liquids, as the only real difference between the two types of fluids is that in liquids the particles are much, much closer together than they are in gases.

Diffusion

You may have heard of diffusion in other contexts, and may even be familiar with it. **Diffusion** is the spread or mixing of a substance from one place to a wider area. When spraying cologne or perfume, we know that the scent does not stay in one location, but moves, and is eventually found in areas that were not sprayed. This is diffusion in action. If you drink coffee or tea with milk, then you have seen as the milk spreads its slowly roiling and billowing tendrils from within the liquid when added. At first, it is only found at the location in the liquid where it was added. Later, the entire beverage is clouded with the milk.

It is not too difficult to see what is happening here. The spritz that is introduced into the air of a room is composed of microscopically tiny droplets that are suspended in the air. (This is termed an "atomized liquid," even though the droplets are much larger than atoms). The droplets start out in a small volume within the room, but each droplet is jostled around as more and more collisions with the constituent particles of the air collide with the atomized liquid droplets. As time goes on, each droplet gets farther and farther away from each other and, therefore, they are spread in a more widely distributed volume within the air. The same thing happens within the liquid volume of the cup of coffee. The milk's individual particles are continually being collided with from the hot liquid particles around it. These milk particles get spread apart from each other and are soon evenly distributed within the coffee or tea. This is diffusion.

Does temperature play a role in diffusion? Yes, it does. You may even want to try this experiment at home for demonstration purposes. Take two separate cups of coffee or tea. Keep one cup very cold and the other on the verge of boiling. Slowly drop a few drops of milk within each at the same time and watch the diffusion process at work. Which cup gets fully clouded first? Most trials should show that it is the hotter cup that diffuses the milk fastest. The heat within the liquid is translated into faster-moving liquid particles. These particles collide with the milk particles with higher energy and, consequently, shove them aside quicker and farther away from each other. Diffusion happens at a faster rate at higher temperatures.

Heat and its Flow

When we think of heat, we usually picture the feeling of warmth, and this is true. Strictly speaking, **heat** is defined as the *transfer* of energy from one system to another. Heat moves. Even though it is easy to do, avoid getting confused between temperature and heat. They are different but similar concepts. Temperature is a measure of the average overall kinetic (moving) energy of the molecules within a substance. Think of temperature as a snapshot photo of the system. It does not move; it is something that is a static property of a material. Heat, on the other hand, can be thought of like a movie. Heat is energy in movement. A material can gain heat or lose heat but never *has* heat, it only has a temperature. Heat and temperature are also different in that they have different units. Just as with any measure of energy, heat is measured in Joules (J), whereas temperature is measured in degrees Celsius (°C) or Kelvin (K) – or, if in the good ole US of A, in degrees Fahrenheit (°F).

If there is an isolated system (no energy comes in or leaves from the system), which is composed of two separate regions of hot and cold, then heat will transfer from the warm to the cold region, until both reach what is called thermal equilibrium. **Thermal equilibrium** occurs when the two areas reach the same temperature.

The movement of heat can occur in several different ways. **Conduction** happens when two bodies are in physical contact with one another. An example would be a car sitting in the Arizona sun, the driver getting in, touches the driving wheel, and gets burned. The high temperature of the wheel and the relatively lower temperature of the driver's hand were not at thermal equilibrium. So, heat from the wheel was *conducted* through to the hand of the driver in an attempt to reach thermal stability. The conduction, in this example, can be explained by the molecules of the wheel vibrating and jiggling at a much higher rate than those in the hand. When they came in contact, the jostling wheel molecules collided with the hand's molecules with such violence as to distribute that energy throughout the hand's molecules. That flow of thermal energy within the hand is the feeling of burning. Metals have a very high thermal conductivity, whereas water, glass, plastic, and air have very low conductivities.

A second way that heat can move around is through a process called convection. **Convection** occurs through fluid substances, such as gases and liquids. Think of the earlier discussion on diffusion where a spray was introduced into a room. That spray eventually spread throughout the room and distributed itself evenly through the process of diffusion. In this case, it is heat, and not molecules, that is diffusing. But, when it comes to heat, it is instead termed convection.

To illustrate in more detail, let's imagine the air molecules in our perfume spray example as liquid molecules within a pot. The spray can be thought of as a localized area of water molecules within the container that happens to be at a higher temperature than the rest of the water. When those faster-moving molecules collide with the slower moving water molecules, they transfer their thermal energy throughout the water until the entire pot of water comes to a thermal equilibrium of even temperature distribution. This spread of heat energy within the liquid water is convection.

The third way of heat transfer is through radiation. **Radiation** is the transfer of thermal energy without the use of a medium to travel through – no physical contact and no fluid medium. Whereas convection occurs through a fluid, and conduction occurs through physical contact between objects, radiation needs neither any fluid nor any contact to move through. Radiation is the only means through which heat can travel through empty space. Astronauts feel the heat of the Sun from the Sun's thermal radiation. This radiation had to travel to the astronaut's bodies through the empty vacuum of space. Radiative heat is also what you feel when holding your hand close to any heat source, without actually touching it, even if all of the air between your hand and the source were evacuated.

If heat is the transfer of thermal energy, and radiation needs no medium to travel through, how does radiation get transferred, and what is the stuff that is being transferred, then? This is a great question and shows the clear delineation between the conductive and convective transfers of heat with that of radiation. Think of radiation as something that is being *radiated*. In the case of visible radiation, it is light that is being emitted. In the case of sunburn radiation, it is ultraviolet rays that are being radiated. In the case of radiographic radiation, it is x-rays that are being emitted. In the case of thermal radiation, it is infrared that is being radiated. All of these different radiations are composed of what are called electromagnetic (EM) waves. That is the "stuff" that moves, and it needs no medium through which to travel. We will learn more about EM waves in the next volume. But at this point, think of them as energy waves that can travel through the void of empty space.

Entropy

Other than the chaotically-inclined, such as children and teens, can it be denied by anyone that a clean room is more orderly than one that has clothes scattered around and not placed in their proper places? Which is in a more orderly state to you: a cup with coffee in your hand or one that has fallen and shattered on the ground? How about between an ice cube or of water from a melted cube – which seems more ordered? Finally, can you distinguish which of these states is the more ordered: a cup of coffee and a container of milk next to it or a cup of coffee with milk that has been poured into the coffee?

What we are seeing is the difference between a state of regularity or order and a state of disorder. The amount of disorder of a system is called its **entropy**, and it plays a significant role in many fields of physics.

However, the order and disorder of a system can be a bit confusing or vague and really depends on the system under discussion. Generally, when talking about entropy, we think about systems that have a lot of pieces or of a lot of particles. A system with trillions of molecules, say, or even a deck of 52 cards, an ice cube with trillions of crystalized molecules bonded to each other, a galaxy with billions of stars, or the universe with hundreds of billions of galaxies; all of these are free game for discussions of entropy.

Let's take the smaller and simpler case of a deck of cards. Arrange the deck so that all the suits are in descending order from highest to lowest rank: ace of clubs, king of clubs, queen of clubs, jack of clubs, 10 of clubs, …, etc., then do the same with the ranks of hearts, the spades, and the diamonds. Obviously, this arrangement of the deck is in a very high order, since there is only one way that the deck can be arranged like this. The deck is now in a state of very *low* entropy ("*low* disorder"). Now, start to shuffle the cards for a while. How orderly will the deck be now? Well, there is now a mix of suits and ranks distributed throughout the pack. It now has a *higher* entropy. More shuffling will only distribute the cards more throughout the pack and, therefore, a more disorderly state of cards will ensue; a higher and higher entropy.

An ice cube is in a lower state of entropy than when found in a puddle of liquid water. The ice is in an orderly crystalline form, where each molecule of water is next to each other in a bonded state. As heat is transferred to the ice from its surroundings, a gradually higher entropy is followed, and the ice begins to melt into a more disorderly mess.

The Zeroth Law of Thermodynamics

Heat follows several central laws of thermodynamics; in fact, there are four such laws. These laws are fundamental in physics, and many physical science and engineering fields make great use of them. Let's discuss each of them in turn.

Imagine three separate and isolated systems A, B, and C. Additionally, within each individual system, they are at thermal equilibrium. There are no more heat transfers occurring *within* each system (due to the reached thermal equilibrium), and the temperatures, as a result, are not changing. Our first law (which happens to be named the "**zeroth law of thermodynamics**," simply because it was the last law to be found but was deemed the most fundamental of the four, and so was placed before the first law) states that:

- if system C is in thermal equilibrium with A and with B, then it follows that A has to also be in thermal equilibrium with B.

The Zeroth Law

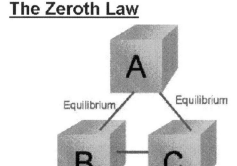

In other words, if one of the systems happens to be in thermal equilibrium with the other two, then those other two systems must also be in thermal stability with each other, despite none of them being in contact with one another.

The zeroth law is pretty self-explanatory and easy to understand. It is the simplest of the four to be stated.

The First Law of Thermodynamics

Our next law of thermodynamics is the most involved and has many far-reaching consequences. Many may already be familiar with it. The **first law of thermodynamics** states:

- The energy within an isolated system is always conserved. The energy within the system can be converted into other forms, but the total amount will always remain the same, both before and after a process happens within that system.

This law is also known as the **law of conservation of energy**. Our isolated system may be a block of solid ice immersed in liquid water, all within a container. The liquid water has a higher temperature and therefore, higher energy than the block of solid ice. The warmer water's heat slowly gets transferred from the liquid water to the ice, and the ice begins to melt. In the end, what we may have left is a container that has nothing but liquid water all at a lower temperature than when it began, but still at a higher temperature than the ice. The energy was conserved. No energy was brought in from outside the system, and none left out of it.

Here's another scenario that demonstrates the first law. Imagine a bicycle turned over and resting on its seat with the wheels in the air. Let's enclose this system within a box. At rest, the bike, the box, and the air in the box, all remain at thermal equilibrium, and no energy is transferred or changed around. Introduce your hand from the outside and reach in and spin one of the wheels, then remove your hand. Let us say this is "State 1".

The wheel spins now, due to the kinetic energy that you've introduced on to it. If you were to add up all of the total energy within State 1, you would come up with a specific value, E. The wheel now has kinetic energy as a result of the spinning. However, being in the real world, this rotating wheel will begin to slow down, and eventually come to a stop. What is happening?

The kinetic energy of the rotating wheel is continuously rubbing against the axle that it is spinning around. The frictional rubbing increases the temperature of that contact area. So, there, the kinetic energy is slowly being converted into heating of the axle. Maybe, as the wheel is turning, there is also a squeaking that is being heard from the wheel and axle. This is a small vibration that occurs at that location, and so some of the kinetic rotary motion of the wheel is not only being converted into heat but now also into vibratory noise. In the end, the wheel is at rest, the axle is a little warmer, the surrounding air is slightly warmer, and the noise has dissipated. We will call this "State 2". When adding the energy of the dissipated noise and the heat, you will have precisely the same energy, E, as what began in the form of the moving wheel. This demonstrates the conservation of energy and the first law of thermodynamics.

The Second Law of Thermodynamics

The **second law of thermodynamics** has to do with entropy. Briefly, it may be stated in a couple of ways:

- The total entropy of the universe increases over time.
- Heat always moves from a hotter body to a colder one, unless some outside work is being done on the system to reverse this process.

Many scientists consider the second law to be the most important in all of science. This is because it has so many far-reaching consequences and explains so much about the universe at large.

The universe, as we now know, is expanding and has, in the past, started out in a much smaller, more compact, and much hotter condition, after its origin in the Big Bang. Initially, the universe was at very low entropy. In other words, it was in a much more orderly state. As time progresses, and the universe expands, it points in an ever-increasing entropic state of disorderliness and decreasing temperature. Indeed, there is more than speculation in physics that this directionality of entropy is precisely what *creates the directionality of time, also known as the "arrow of time,"* itself! As the late, great physicist Stephen Hawking had said:

> *"The increase of disorder or entropy with time is one example of what is called an arrow of time, something that distinguishes the past from the future, giving a direction to time."*

Why does entropy tend to increase over time, instead of decrease? In other words, why does disorder tend toward an increase, instead of a reduction, over time? To explain this, we need to go back to our deck of cards. When the cards were placed in the original arrangement of order, we needed to do work on the system of the deck of cards to get them into this state. If, however, we start to shuffle the deck, the cards will move around in a somewhat random way. There are so many more arrangements that the cards can be in that accord them a disorderliness than the one that finds them in the original arrangement of ordered suits and ranks. Let's explore this idea further.

So, you have a deck of 52 individually distinct cards; each card being unique. Now, how many different arrangements of a single pack of 52 unique cards are there? Well, the first card's position can be occupied by one of any of the 52 cards in the deck. The second card in the arrangement can be one of any of the remaining 51 cards. The third card in the arrangement can be one of any of the remaining 50 cards, and so on, until the very last card is placed into the 52nd place. How is this determined mathematically?

Let us look at a much smaller deck of only four cards to illustrate the math. The first place in the deck can be taken up by any one of the 4 initial cards. The 2nd place can be taken up by any of the remaining 3 cards, etc. So, the way this is calculated is by multiplying each of these places by the number of possible cards remaining for that place. So, in the case of the deck of 4 cards, this amounts to multiplying 4 X 3 X 2 X 1 = 24. There are 24 different ways that 4 cards can be shuffled together, or 24 different and unique arrangements of cards. When multiplying a number incrementally like this, by starting with the highest number and going down to 1, this is called a *factorial* and is represented with an exclamation point, like so: 4! = 4 X 3 X 2 X 1.

Let's add another card to the small deck of 4 cards, to make 5 total cards. We may ask how many different arrangements can *this* 5-card deck be in? Well, it will be "5!" or 5 X 4 X 3 X 2 X 1 = 120 different arrangements. Here's a table of the first few numbers of small card deck arrangements:

Number of Cards in the Deck	Number of Different Arrangements
4	4! = 24
5	5! = 120
6	6! = 720
7	7! = 5,040
8	8! = 40,320
9	9! = 362,880
10	10! = 3,628,800

Notice how quickly the number of possible arrangements goes up as a new card is added. Going back to our original deck of 52 cards. We do the same thing, but this time there will need to be 52 multiples in the equation: 52 X 51 X 50 X...X 1. Or, writing in mathematical symbolism: 52! ("52 factorial"). What does this equal? Hold on to your seat, because it is more than enormous:

$$52! = 8.07 \text{ X } 10^{67}$$

That is, the number 8 followed by 67 zeros. This represents how many different and unique arrangements that a single deck of 52 cards can be made into. Let's make this number a tiny bit clearer. This far exceeds the number of atoms that make up the entire earth (which is estimated to be about 10^{50} atoms)! If you were to put a deck of cards in a uniquely different arrangement, once every second, *and* started doing this at the very beginning of the universe, about 14 billion years ago...you will still be arranging this deck many more millions of years into our future! This is not a small number.

This goes without saying, but this means that every arrangement of cards you will ever see in your entire life will be unique, and will probably never be seen again. That is *unless* it is physically and intentionally moved into a different arrangement. This brings us back to the well-ordered arrangement of suits and ranks that we discussed before. There is only one unique arrangement that that well-ordered arrangement can be in. When shuffling, there is only one chance in 8.07×10^{67} that this will ever occur again, but there is *every other* chance that the arrangement of cards will be in a different setup.

The same description happens at the atomic scale of molecules and with stars and galaxies when we are talking about arrangements or orientations of multiple objects. The ice cube consists of trillions of nicely arranged water molecules. However, apply some heat to it (which is equivalent to shuffling that deck of cards), and that arrangement gets effectively destroyed. The same can be said of the initially ordered arrangement of energy and matter in the early universe. The beginning of the universe at the Big Bang was the most ordered arrangement of matter and therefore the lowest entropy state. When the universe started to expand afterward, there was more space available for this matter and energy to arrange themselves or to be located in. Therefore, a higher number of disordered arrangements were available for this material, and so they were more likely to be found in them. Thus, entropy naturally moves from a lower state to a higher one.

But how does this work with heat? Why does a system tend to move from hot to cold until a thermal equilibrium among its parts is found at the end? Imagine a box that is initially set up with a separating partition in the middle. Thousands of fast-moving (hotter) molecules on the left side of the box are separated from much slower moving (colder) particles on the right side. The left side is much warmer than the right side due to those faster-moving particles colliding with each other and with the walls at higher rates and with more energy. This state is one with *low* entropy because there is a specially ordered arrangement of hot and cold molecules. Now, open the partition between the two sides. How likely is it that this partitioned arrangement will continue while the barrier is removed? Not likely at all, because now particles from both sides can fly around the entire box and are no longer constrained. Many faster particles will eventually fly to the right side, and many of the slower molecules will find themselves on the left side of the box. They will mix and collide with one another, imparting kinetic energy from faster moving particles to the slower particles until an average is reached. What this means to us is that the higher temperature left side will gradually warm the right side, but by doing so, the left side will slowly cool. There is a heat exchange where heat moves from left to right until both sides are at the same temperature, and thermal equilibrium is found. You will never find a cold region spontaneously turn hot. There is just so much more space for the bouncing hot, and cold particles to now find themselves in, that they will randomly mix both sides together of their own accord. Entropy, in this case, moves in the direction from temperature difference to thermal equilibrium and increases in that direction. From a hotter, early universe to a cooler, later universe. This is the second law of thermodynamics.

The Third Law of Thermodynamics

We finally meet our last law of heat, the **third law of thermodynamics**. The third law states:

- As a substance's temperature approaches absolute zero, its entropy also approaches zero.

Remember that the temperature is a measure of the average kinetic energy of a substance; it measures how much jiggling or motion there is within the material. So, as its temperature lowers, there is less and less motion occurring. There comes a theoretical point in the Kelvin scale where no movement happens at all. I say "theoretical" because, in reality, it is not even possible to reach this point, called **absolute zero**. You can approach it but never reach it. The reason why this is so is due to the second law of thermodynamics. Areas are always surrounded by at least some sort of ambient heat that can be transferred to other locations. So, when an area nearby becomes colder than an adjacent one, the adjacent area's heat will spontaneously move to the colder area to reach thermal equilibrium. There is also a quantum mechanical component to why absolute zero can never be reached, but that discussion will have to be made elsewhere or elsewhen.

Absolute zero is -273.15°C (-459.67°F). Experimentally, we have come very, very close to this temperature. As of this writing, the United States laboratory of the National Institute of Standards and Technology (NIST) in Boulder, Colorado has created the lowest temperature ever recorded in 2018 - 2019. They lowered the temperature of a gas that consisted of the elements rubidium and potassium down to within 50 billionths of a Kelvin above absolute zero, or 50 nanoKelvin (50nK)! Of course, at these extremely low temperatures, the substance being observed will be in a solid phase.

AFTERWORD

We have arrived at the end of our look at thermodynamics and to the end of this volume of physics. In this volume, we had focused our attention on the study of the more familiar aspects of physics.

This included classical mechanics, where we looked at the laws of motion and the many ways in which objects move. We saw objects move linearly in constant and accelerating motions, in spinning angular movements, and in free falls. We analyzed masses and their properties under gravity, impacts, forces, heat, and cold. Our studies brought us to the world of energy, both kinetic and potential, and its uses. We did work and performed power.

Vibrations and waves were studied and described. Sound, sonar, and the study of seismology were explained.

Then, we made our way into the world of fluid dynamics, where we looked at liquids and gases. We moved on to temperature differences and how to measure them. Finally, we reached the science of thermodynamics, where we discussed its importance for many other fields of physics, and also to the origin and fate of the universe.

We traveled back in time to ancient Greece and Medieval Europe, and back again to modern times.

None of our discussions would have been complete without rounding it all off with looking at many of the laws of physics. These included Newton's all-important three laws of motion and his law of universal gravitation. It also included the conservation laws of linear momentum, angular momentum, and mechanical energy. Finally, we discussed all of the four laws of thermodynamics, including entropy.

However, this is by no means the conclusion of our study of physics. There will be two more volumes. Our second volume will begin studying the less familiar, but just as important, fields of optics, electricity, and magnetism. The third volume will delve into the very topics of physics that I had waited in great anticipation to study when I was majoring in it - relativity theory, quantum mechanics, and others.

I thank you, dear reader, for your generous time and patience in following me through this diverse subject. It is an absolutely fascinating one and is also very dear to my heart. This has been my labor of love. It should give us all some pause in what we now know, and in how much more appreciation and awe we should now have for the very home we find ourselves in – our universe.

APPENDICES

APPENDIX A: Kepler's Laws of Planetary Motion

APPENDIX B: Weightlessness in Space and Orbits

APPENDIX A
KEPLER'S LAWS OF PLANETARY MOTION

Johannes Kepler

Johannes Kepler (1571-1630) was born in Weil der Stadt, Germany to a mercenary father, who died when Kepler was young, and to a healer and herbalist mother. Kepler attended the University of Tübingen and studied theology and philosophy there. His original intent was to become a minister. Recognized by his greatness in mathematics at the university, Kepler applied this to his cosmological understanding. During his studies, he learned about both the Ptolemaic and Copernican planetary motions and attached himself to the sun-centered Copernican model.

Later, Kepler would hold a teaching position at a Protestant school in Graz, where he also met and married Barbara Müller. They had five children together, two of whom died in infancy. While teaching in Graz, he would write his first major work, *Mysterium Cosmographicum*, which was the first published work that argued in defense of the Copernican system.

Kepler's next move was to Prague, where he worked with the Danish astronomer Tycho Brahe. At the time, Tycho's precision planetary data was the best in the world and was taken from his own telescopic observatory. Kepler was working on his own model of the solar system, but to pin down its accuracy, he needed access to Brahe's astronomical data. Brahe, however, would only feed Kepler tidbits of this information. Because of this, Kepler became impatient with Tycho and formed an uncomfortable and tense relationship with him.

In 1601, Tycho died, and Kepler inherited his title of Imperial Mathematician, where he gave astrological advice to Emperor Rudolph II. With Tycho's newly inherited data, he was able to formulate his first two laws of planetary motion, and subsequently published them in his 1609 work, *Astronomia Nova*.

In 1612, the Lutherans were forced out of Prague, due to the counter-Reformation. Tragically, Kepler's wife, Barbara, and two sons died, and he eventually moved on to Linz. He continued teaching in Linz, but soon had another misfortune when he had the responsibility to defend his mother, Katharina, from accusations of witchcraft. In 1613, Kepler married his second wife, Susanna Reuttinger. With Susanna, Kepler had a much happier marriage, but tragedy was to strike yet again. His first three, of six, children to her were to die in their childhood.

In 1619, he published *Harmonices Mundi*, "Harmony of the World." In this work, Kepler laid out his third law of planetary motion. *Harmonices Mundi* attempted to explain how the orbits of the then-known six planets (Mercury, Venus, Earth, Mars, Jupiter, and Saturn) could fit within celestial versions of the five perfect Platonic solids (the octahedron, icosahedron, dodecahedron, tetrahedron, and the cube). This model was found to be incorrect, and Kepler had to, depressingly but bravely, dismiss it.

Kepler's last work was considered during his time to be his most significant work. This work, *Rudolphine Tables*, was published in 1627. Named after the Holy Roman Emperor, Rudolph II, it was a catalog of stars and planetary data, much of which was the original data accumulated and tabulated by Tycho Brahe.

On November 15, 1630, Johannes Kepler died in Regensburg, Germany, where he was buried.

Kepler's First Law of Planetary Motion

Before Kepler's laws of planetary motion, the accepted Copernican system assumed that the planets orbited the Sun in perfect circles with the Sun at the very center. This also meant that the orbital speed of the planet in its orbit was always constant as well.

After examining and modeling Tycho Brahe's data into more accurate orbital trajectories, Kepler found that the planets did not actually orbit the Sun in circles. This led him to propose his first law:

An **ellipse** is a geometric figure in the shape of a squashed circle. A circle has a single center, but an ellipse has two *foci*. An ellipse is created by first placing two separated, but fixed points, such as two nails, on a slab of wood. We call these points the *foci*. Attach one end of a piece of string onto one of the nails and the other end of the line to the second nail, with some slack to the string. Hold a pencil or a pen snugly against the inside of the string and begin writing an ovoid shape around the nails. The shape that will be produced is an ellipse.

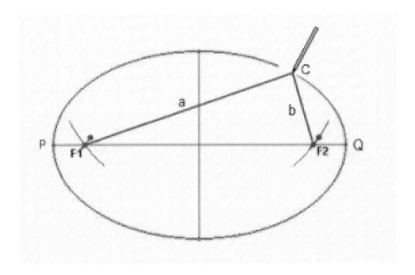

In the figure above, F1 and F2 represent the foci. The interesting thing about an ellipse, and the defining characteristic of one, is the fact the total of length "a" (distance from focus F1

to the point "c" on the ellipse) and length "b" (the distance from focus F2 to point "c"), will always remain the same, no matter where point "c" is on the ellipse.

What Kepler found was that the Sun is located at one of these foci and each planet has an orbit that is an ellipse around the Sun, not a circle like earlier models assumed. To be sure, these planetary ellipses were not very far off from actual circles themselves, but far enough to make a difference in the data. Kepler's first law changed from the earlier Copernican model in three critical ways:

1) The planets no longer orbited in circles but in ellipses,
2) The Sun was no longer at the very center of the solar system but at one of the focal points,
3) The planets did not move in their orbits at constant speeds but changed throughout their orbits, which leads us to Kepler's Second Law of Planetary Motion:

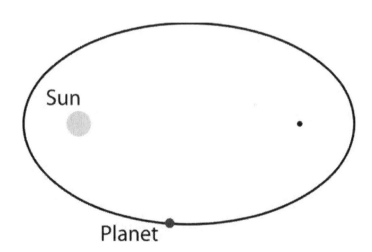

Kepler's Second Law of Planetary Motion

Given his first law of planetary motion, Kepler found that there was additional interesting information inherent within it. This information also closely followed the data that Kepler found from Tycho Brahe's collections of planetary motions across the sky.

What Kepler discovered was that each planet in its orbit around the Sun could not move along it at a constant speed, but a varying one instead. The planet moves in an ellipse with the Sun at one focus and nothing found at the other focus. As the planet in its orbit approaches closer to the Sun, it will pick up more speed and sweep by the Sun. As the planet moves farther away from the Sun, it gradually slows down. It repeats this process throughout its orbit.

Kepler found a very interesting geometric equality in his study of this movement in the planet's orbit. This brings us to Kepler's Second Law of Planetary Motion:

> *A planet sweeps out an equal area in equal times around its elliptical orbit.*

Let's explain this a little further. In the figure below, you can see two sectors that are shaded black and grey. These are equal areas. When the planet moves along its orbit its speed changes depending on where in the orbit it is located. So, when the planet is farther from the Sun, and it moves from point X to point Y, it takes the same amount of time as it would to move from Point A to point B when closer to the Sun. There is a swept-out pie-shaped sector that is created as the planet moves. The black swept-out area is precisely the same amount of area as that swept-out in grey. While the black is closer to the Sun, fatter and wider, the grey is farther, thinner and more stretched out, but still the same amount of areas.

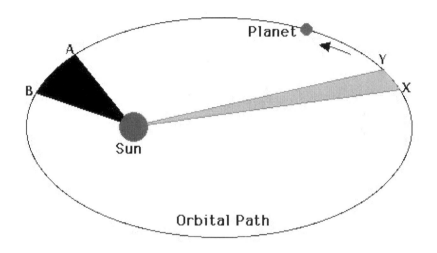

Kepler's Third Law of Planetary Motion

Kepler knew, from his collected planetary data, how far away each planet was from the Sun and how fast each world moved at every point in its orbital path. From this information, he was able to extract yet another pattern that became known as his third law of planetary motion. As this law is a thoroughly mathematical one, we will need a couple more pieces of knowledge to fully understand it.

If you were to extend an imaginary line through the center of an ellipse and pass it through the two foci, this line would be

215

called the **major axis**. Split the line into two equal halves, and each one is then called a **semi-major axis**. We'll call distance this "R," for short. The distance of earth's semi-major axis is about 93 million miles (this just happens to be the average distance that Earth is from the Sun in its orbit). But, because the Earth is a very special planet for us, we have a unique name given to this distance. We call it the **astronomical unit**, or AU for short. Earth is 1 AU away from our Sun (R = 1 AU).

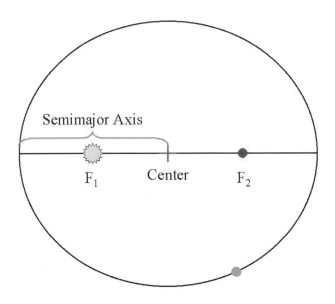

One last piece of information we need to be familiar with. The time it takes for a planet to orbit the Sun in one go-around is called that planet's **orbital period**, or "T" in math symbolism. Earth's orbital period is 365.25 days.

Now, for the payoff. When Kepler was tabulating his data, and making comparisons between all the known planets of his time, he was awestruck by an amazing mathematical find. When he cubed any planet's semi-major axis, R^3, and then squared the same planet's orbital period, T^2, and lastly, took their ratio, he found that it always resulted in the same number:

$$\frac{R^3}{T^2} = 7.5$$

In words, this is Kepler's third law of planetary motion:

The square of a planet's orbital period is directly proportional to the cube of the semi-major axis of its orbit.

His third law was described in his *Harmonices Mundi*. In it, Kepler writes of his excitement at discovering this law:

"I first believed I was dreaming... But it is absolutely certain and exact that the ratio which exists between the period times of any two planets is precisely the ratio of the 3/2th power of the mean distance."

APPENDIX B
WEIGHTLESSNESS IN SPACE AND ORBITS

The wizards of outer space

We have all seen those bizarre videos of astronauts (or "cosmonauts," if they are from Russia) floating around within their spacecraft, playing with food and drink, and performing unearthly acrobatics. I fondly remember big-screen IMAX theater documentaries of the space shuttle and ISS (International Space Station) astronauts performing their EVAs (Extravehicular Activities) outside of the shuttle or station and the viewers going along on the rides with them. For us, earthly surface-crawlers, these activities had the playful name of "spacewalking."

Fellow astronauts would help their comrades suit up and attach their space helmets to the suit. They would follow through with checklists to verify that everything on their uniforms was ready for the vacuum of outer space and that no rips, tears, or defects existed on them. Any of these could spell disaster to the brave souls within the essentially life-support bags we call spacesuits that surrounded the astronauts. They were ready for their EVA.

The astronauts would be escorted through a cramped little tunnel within the craft, and their unsuited fellows would stay behind. A door would close behind the astronauts, and they would now be hermetically sealed from the rest of the ship. This corridor would soon become a vacuum chamber, where all of the air within escaped to the emptiness of space. The suited astronaut's next step was to undo the seals around the outer door and unlock it. A brief air sucking sound was heard and then vanished. The only remaining sound was their own breathing within their helmets. Swinging the door open, a flood of sunlight entered the chamber, and the astronauts were now free to perform their EVAs.

Sometimes EVAs include repairing a portion of the outside of their space vehicle. Maybe some of the heat-resistant hard foamy tiles on the underbelly of the shuttle needed replacement or just a visual scan of the entire ship was required. Another required activity during the spacewalk may be to coordinate with other fellow spacewalkers to move a communications satellite out of the cargo bay of the space shuttle to prepare it to go into orbit around the earth for use. Perhaps the Hubble Space Telescope needed to be docked with to replace one of its faulty mirrors.

Gargantuan satellites or telescopes, that would dwarf the astronauts, would be gently moved about by the space-suited "Merlins" as if by implementing a hovering magic spell.

How did all of this magic happen? How do the astronauts feel this weightlessness in space, and how can it be that they can manipulate large masses and move them around so freely?

I can sum the answer up in two words: free fall. However, it will take a little explaining to fully understand it.

Desperately Seeking Zero-g

A strange feeling comes over us when we plummet downwards, all of a sudden. Take the gentle and controlled fall down several floors in a hotel elevator. We've all had that strange twisting or tickling feeling within our stomachs when it happens. Our sense of orientation with standing upward and stable becomes momentarily disrupted, and our body floods us with a quick rush. But it quickly passes, and our downward journey resumes unnoticed.

But the elevator is mild. Instead, take the next adventurous step and get buckled up within a roller coaster. As that carriage you happen to find yourself in inches its way to the very peak of the rails, you slowly get a glimpse of the surrounding world below and around you. The ground is far below, and your body's muscles tense up with the agonizing anticipation that your entire body is about to experience a great fall. And you do. And you scream.

There are those in society who feel that the roller coaster is but child's play and that they would rather have their dear lives hang, literally, by a thread…or less. These people are what some may call mad. But this madness is all too pervasive, and there is a sort of cottage industry that is sustained by their dollars. These adventurers go the route of bungee-jumping, skydiving, and wingsuit flying, only one category of which is hanging by a thread.

You may have thought that the previously mentioned category of people was the worst of the bunch when it comes to seeking dangerously falling activities, but you would be wrong. There is a category of bravery that has found the ultimate in falling, and they are called astronauts.

Essentially, what all of these people are pursuing is something called "zero-g." Free fall is one of the methods you generate it. On the good, safe, and solid surface of the Earth, we are captives of the ground underneath us. The earth pulls us to it. Our mass feels this pull. However, if there was not something under us also pushing us *up*, where would we go? We would continue falling through the ground! So, the ground is pushing up on us. This is called a *normal force*. We feel this pushing up of the ground as our heaviness or weight. The more the ground is pushing against us, the heavier we feel.

However, you may think, our weight is pointing *down*, not up! This is correct. The earth's gravity points downward, pulls on us, and creates our weight. Remember, from chapter 4, that weight equals mass times the acceleration of gravity ($W = mg$). Also, remember normal forces and Newton's third law that says that every action has an equal and opposite reaction. Your weight is pushing down on the earth's surface. That's the action. The Earth must now react, by pushing up on your feet, with the same amount of force, to prevent you from falling through the ground and to hold you up. That's the normal force, and it is *that* force that you feel.

To prove this, we can use a weighing scale that is used to measure your body weight. You will need the scale brought into an elevator. (The best scale is the analog kind with the moving pointer). Don't worry, just tell your inquisitively-looking fellow passengers that you are performing science! Place the scale on the floor within the elevator and stand on it. Your weight should show correctly as usual. Now, press one of the other floor buttons and observe what happens to the scale. If you are moving down some floors, the weight will point lower. On the other hand, if you are moving up some floors, the pointer will move higher, registering greater weight.

Can this be true? Is your weight actually decreasing and increasing before your eyes? Yes, it is! And it is that normal force from the scale pushing up on you that is doing it. It is precisely the same as if you were to hold the device in your hands and were to push it against the side of your body. Obviously, you are not standing on it, but the scale is still measuring the force that it pushes against you. When the elevator moves down some floors, the scale applies *less* force upward against your shoes. You feel lighter. Whereas, when moving upward, the device applies *more* force against your shoes. You then feel heavier.

So, how does all of this relate to free fall? Well, if that elevator's cables were to break and come loose (*just a thought experiment!*), the elevator, the scale, and you within it will all fall freely. As this was happening, and if you were to be standing on the weight scale and watching the pointer, you would see it move to the zero mark. No weight would be measured. This is called free fall, and you would be in a state called **zero-g**. Zero gravity. Gravity is what gives us weight. If gravity were to be dialed down to nothing, then we would have none. That is called zero-g. We just happened to create it artificially through the use of a falling elevator (or a bungee jump, a wingsuit fall, or a skydive).

Principle of Equivalence

This may seem very counterintuitive, as the Earth still has gravity and you are near the Earth during these falls, why would you not also feel earth's gravity while falling? Albert Einstein made a similar thought experiment himself. Only he did it in the opposite way, and he discovered the answer.

When Einstein was working on his General Theory of Relativity, he imagined someone sleeping in a box with no windows, like an elevator, that happened to be floating in space. Now, if, on the outside, a continuous force was applied to the underside of the box, the box would accelerate upward, and the person inside would be pushed against the ground. If that force happened to be strong enough to move the box at an acceleration of about 9.8 m/s^2, then the person would feel "one-g," or the same as on earth's surface. As the box is accelerated, the person wakes up from their slumber. How would the person know if they were really feeling earth's gravity or if they were accelerating upward? They wouldn't. He found equivalence between the force of gravity and acceleration and called it the **Principle of Equivalence**.

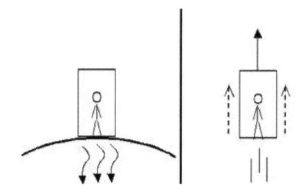

Principle of Equivalence. The left illustration shows a person experiencing gravity from a planet. In the right figure, the person is experiencing the same sensation, but the source is from the acceleration of the elevator.

In other words, Einstein found that the force of gravity is exactly the same as an acceleration of a mass. The feeling is even the same. So, if you were on that elevator again, without moving it, and it happens to be in a gravitational field, like near earth, the person would feel their usual weighty self. They would feel the same as being on earth. But reversing Einstein's thought experiment, by having the elevator accelerate *towards* the Earth, will diminish and cancel the feeling of gravity to zero. This is zero-g, and this is how Einstein explained the localized loss of gravity (during free fall) in a gravitational field.

Orbiting Wizards

Okay, let us go back to our EVA astronauts. They are all experiencing zero-g in their activities in space. The earth is below them, so obviously there has to be gravity. The astronauts are floating so they have to be in zero-g. Additionally, they do not seem to be falling towards the earth. What gives? The astronauts are free-falling, like in the elevator. But then why do they not fall all the way down to the Earth? Because there is one more movement that they are making that most people forget about, and that makes all the difference in the world. All of them (and their spacecraft) are also moving *forward* perpendicularly to the surface of the earth at a very high rate of velocity.

As most of us know (and hopefully, believe), the earth is round. Let's assume you are on the surface of the Earth and aim an unbelievably powerful cannon towards the horizon. Let us also assume that the projectile does not feel any force of gravity; it just shoots straight ahead in a straight line forever. The question is, what happens to the ground underneath the projectile as it flies forward? Can you see that the ground will appear to gradually drift farther and farther away? This is because we know that the earth is curved. The fascinating thing about this curving away of the planet is that it *accelerates away*. The ground will recede away slowly at first, but then at a quicker and quicker rate, not just at a steady constant rate.

No gravity Scenario

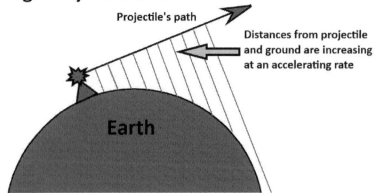

No gravity projectile path. The ground will appear to recede away from a projectile at an accelerating rate.

Now, let's turn on the pulling force of gravity. The cannon shell will start to drop, correct? It will fall faster and faster toward the ground, due to the acceleration of gravity. But if the accelerating ground does not move away fast enough, the falling of the shell will be quicker, and the projectile will eventually hit the ground. If, however, the shell's rate of fall is *slower* than the ground's acceleration away from it, then the shell will eventually move farther and farther away from earth and out into space.

A unique situation occurs when both ground and projectile fall rates are the same. This is when an **orbit** happens. There will be a perpetual falling of the projectile and a ground that is continuously moving away at the same speeds. The missile will remain at the same height above the earth and will be in orbit around it.

To help illustrate this concept of an orbit better, there is a wonderful diagram I remember seeing in Isaac Newton's 1728

De Mundi Systemate (*"A Treatise of the System of the World"*). Here it is in all of its glory:

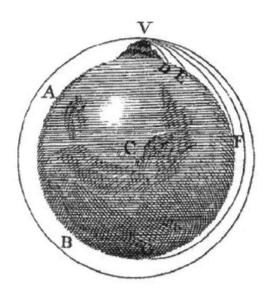

As you can see, there is an absurdly high mountain on the earth labeled "V." An imaginary cannon is erected on the peak and shoots increasingly powerful projectiles to the right. The trajectories of all the different shots fired are labeled with their respective curved lines. The shot fired labeled "D" was with little power, and it eventually fell close to the foot of the mountain. Shot "E" fell a little farther away because it had a little more initial velocity in the rightward direction when shot from the mountain. The same thing can be said of shot "F," but this time it was fired with still more power. In fact, shot "F" was fired with such force as to just about clear the horizon at the peak and fall over it slightly. Shot "B" lands on the other side of the Earth from where the cannon is located. Finally, shot "A" was fired with the power to keep the projectile in orbit altogether. It may have even hit the backside of the mountain itself, judging from the diagram!

What you would see if you were on each projectile is the ground coming up to you at a slower and slower rate with each shot, until finally with shot A the ground remains at a constant distance away from you.

The important thing to note with all of these illustrations is the fact that *the **same rate of free fall is happening with each and every one of these scenarios***. The projectiles all fall to the ground at precisely the same rate of acceleration. It is just that the ground may fall away from the missiles at different speeds until projectile A's fall cannot keep up with that of the ground's fall.

So, our celestial wizards and their spacecraft are all falling at the same rate of free fall as the earth's surface accelerates away from them and keeps them in orbit around the planet. Since they are freely falling, they are also experiencing zero-g to boot. Thus, ends our story of astronauts, spacecraft, orbits, zero-g, and magic wands.

INDEX

- **Bold** index entries are those that have definitions associated with them or are within sections devoted to their topic.

ABOUT THE AUTHOR

Frank Maybusher holds a degree in physics and a minor in mathematics from the University of South Florida. For his undergraduate research in physics, he designed and developed a fully automated radio solar tracking observatory, which was located on the roof of the university's physics and mathematics department building.

Later in his academic career, Frank Maybusher performed research in graduate astrophysics and cosmology at Arizona State University's School of Earth and Space Exploration. Some of this research included making observation runs of extremely red-shifted objects at the Fred Whipple Observatory on Mount Hopkins, Arizona. These objects represent quasars at the farthest edge of the observable universe, some of which were seen for the very first time in this research.

Additional research at ASU included creating cosmological modeling of galactic superclusters and voids using the IDL software language.

Frank Maybusher's career path has been a very eclectic one, finding himself working with a security clearance at Honeywell Aerospace in Clearwater, Florida. He helped develop and test navigational ring laser gyros for military, aerospace, and satellite applications. While in Florida, Maybusher also worked in industrial labs stress-testing and examining turbojet engine parts using CCD video and microscopic analyses.

Later, he worked for companies as a software and database engineer.

Frank Maybusher has also lived an alternate life in the dental field. He has received a Doctor of Dental Medicine (DMD) degree from the University of Louisville, School of

Dentistry and is very happily working as a dentist with a great and loving team.

Outside of work, Dr. Maybusher has enjoyed SCUBA diving, hiking, and traveling the world. He is a member of MENSA, the Triple Nine Society, and the International Society for Philosophical Enquiry. Dr. Maybusher is very happily, and lovingly, married to his wife, Regina. They both have two precious sons, Isaac and Aeon. Both of whom Dr. Maybusher hopes, *with all of his heart*, will see the light and eventually pursue careers in science. Maybe they, too, will be touched by the true beauty and glory that the universe holds, if only their eyes were to look up!

Made in the USA
Columbia, SC
19 January 2022

53928408R00130